# Chlorodioxins—Origin and Fate

# Chlorodioxins—Origin and Fate

**Etcyl H. Blair,** *Editor*

A symposium sponsored by the
Division of Pesticide Chemistry
at the 162nd Meeting of the
American Chemical Society,
Washington, D. C.,
Sept. 16-17, 1971.

ADVANCES IN CHEMISTRY SERIES **120**

AMERICAN CHEMICAL SOCIETY

WASHINGTON, D. C. 1973

ADCSAJ 120 1–140 (1973)



Library of Congress Catalog Card 73–84139

ISBN 8412–0181–1

PRINTED IN THE UNITED STATES OF AMERICA

# FOREWORD

ADVANCES IN CHEMISTRY SERIES was founded in 1949 by the American Chemical Society as an outlet for symposia and collections of data in special areas of topical interest that could not be accommodated in the Society's journals. It provides a medium for symposia that would otherwise be fragmented, their papers distributed among several journals or not published at all. Papers are referred critically according to ACS editorial standards and receive the careful attention and processing characteristic of ACS publications. Papers published in ADVANCES IN CHEMISTRY SERIES are original contributions not published elsewhere in whole or major part and include reports of research as well as reviews since symposia may embrace both types of presentation.

# CONTENTS

# PREFACE

Chlorinated dibenzo-*p*-dioxins long have been recognized as possible by-products in manufacturing certain chlorinated phenols (*1*). Current interest in chlorinated dibenzo-*p*-dioxins has originated because the highly toxic 2,3,7,8-tetrachlorodibenzo-*p*-dioxin appeared in trace elements in some samples of the herbicide 2,4,5-trichlorophenoxyacetic acid (2,4,5-T).

The initial concern for the possible hazard to humans exposed to 2,4,5-T was precipitated by teratologic studies conducted by Bionetics Research Institute under contract from The National Cancer Institute (*2*). In these studies, large doses of 2,4,5-T were administered to pregnant rats and mice for nine of the 21 days of pregnancy. The incidence of fetal abnormalities was slightly higher in the treated animals than in control animals. Later tests indicated that these abnormalities (cleft palate) may have been caused by 27 ± 8 ppm of 2,3,7,8-tetrachlorodibenzo-*p*-dioxin present as a contaminant in the 2,4,5-T sample used in the Bionetic study (*3*). After the results of the study were made known, the Panel on Herbicides of the President's Science Advisory Committee studied the total 2,4,5-T situation. The report of this committee was published in March, 1971 (*4*).

In addition to its extremely high oral toxicity, skin contact with substances containing 2,3,7,8-tetrachlorodibenzo-*p*-dioxin may allow toxicity in the form of chloracne, a condition characterized by eruptions of the skin on the face, neck, and back. Also, chlorinated dibenzo-*p*-dioxins have been associated with the "chick edema factor," a disease of chicks associated with contaminated fats or oils used in the manufacture of their feed.

Since 1950 many in the chemical industry have been keenly aware of the possibilities of highly toxic 2,3,7,8-tetrachlorodibenzo-*p*-dioxin forming in the manufacture of 2,4,5-trichlorophenol. This phenol is made by treating 1,2,4,5-tetrachlorobenzene with strong caustic at high temperatures for several hours. Manufacturing procedures which do not carefully control temperature and alkalinity increase the probability of dioxin formation.

Higher chlorinated dibenzo-*p*-dioxins, such as the octachloro derivative, can be formed in manufacturing pentachlorophenol from hexachlorobenzene. However, this highly chlorinated dioxin is remarkably different in chemical, physical, and biological properties from that of the 2,3,7,8-tetrachlorodibenzo-*p*-dioxin and is much less toxic. Theoretically, 75 isomers of chlorinated dibenzo-*p*-dioxins are possible, but few have been synthesized and studied in detail.

This volume should do much to advance the knowledge of these types of molecules, and additional research should be stimulated to explain further the chemical, physical, and biological properties of the chlorinated dibenzo-*p*-dioxins.

### *Literature Cited*


1. Bauer, H., Schultz, K. H., Schultz, V., Spiegelberg, X. X., *Arch. Gewerbepathol. Gewerbehyg.* (1961) **18,** 538.
2. Higginbotham, G. R., Huang, A., Firestone, D., Verett, J., Ress, J., Campbell, A. D., *Nature* (1968) **220,** 702.
3. "Effects of 2,4,5-T on Man and The Environment," Hearings Before The Subcommittee on Energy, Natural Resources, and The Environment, April 7, 15, 1970; Serial 91-60, pp. 375-376.
4. "Report on 2,4,5-T," A Report of The Panel on Herbicides of The President's Science Advisory Committee, Executive Office of The President. Office of Science and Technology.



Etcyl H. Blair

Midland, Mich.

February 1973

# 1

# The Preparation of Uniformly Labeled $^{14}C$-2,7-Dichlorodibenzo-*p*-dioxin and $^{14}C$-2,3,7,8-Tetrachlorodibenzo-*p*-dioxin

WESLEY W. MUELDER and LEWIS A. SHADOFF

Ag-Organics Department and Analytical Laboratories, The Dow Chemical Co., Midland, Mich. 48640

*2,7-Dichlorodibenzo-*p*-dioxin was prepared from isotopic potassium 2,4-dichlorophenate uniformly labeled with $^{14}C$. Ullman conditions gave a 20.5% yield. Small amounts of dichlorophenoxy chlorophenol were removed from the product by extraction with sodium hydroxide before purification by fractional sublimation and recrystallization from anisole. Chlorination of 2,7-dichlorodibenzo-*p*-dioxin in chloroform solution containing trace amounts of $FeCl_3$ and $I_2$ yielded a mixture of tri-, tetra-, and pentachloro substitution products. Purification by digestion in boiling chloroform, fractional sublimation, and recrystallization from anisole was effective in refining this product to 92% 2,3,7,8-tetrachloro isomer, which also contained 7% of the tri- and 1% of the penta-substituted dibenzo-*p*-dioxin. Mass spectroscopy was used exclusively to monitor the quality of the products during the synthesis.*

The identification of 2,3,7,8-tetrachlorodibenzo-*p*-dioxin (TCDD) as one of the active acnegens in commercial 2,4,5-trichlorophenol was reported in 1957 by Kimmig and Schulz (*1*). That it is formed during the alkaline hydrolysis of tetrachlorobenzene was supported by Jones and Krizek (*2*), who found that TCDD is produced when sodium 2,4,5-trichlorophenate is held at 350°–400°C for 30 hours. Because of its acute toxicity to some laboratory animals (*3*), this material was handled with extreme care.

Currently, there is considerable interest in the effects of this chemical on the ecology (*4*) as it is deployed through using certain herbicidal formulations (5). Studies (6, 7) concerned with its fate in the environment were designed, using a radioactive tracer, a technique which often furnishes information very difficult or impossible to obtain by other methods. Therefore, the synthesis of uniformly labeled $^{14}C$-TCDD was undertaken, and its preparation is reported here.

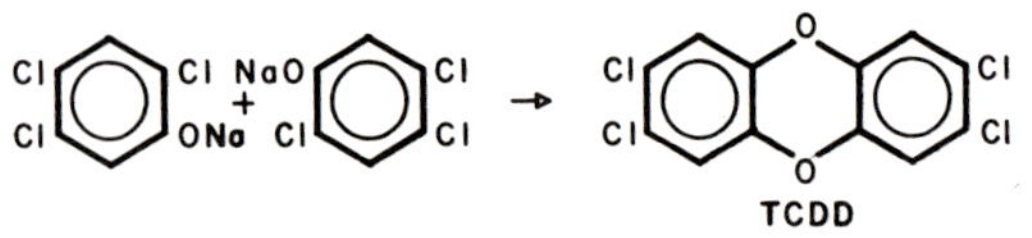


*Figure 1. Preparation of 2,3,7,8-tetrachlorodibenzo-*p*-dioxin from sodium 2,4,5-trichlorophenate*

In selecting the route for the isotopic synthesis, two methods were considered. Pyrolysis of sodium 2,4,5-trichlorophenate (2) or its condensation under Ullman conditions (8) (Figure 1) would directly yield TCDD. Exotherms observed in the pyrolysis of higher chlorinated phenols (9) and relatively low yields made the pyrolysis method unattractive. The less strenuous conditions of the Ullman reaction, as modified by Aniline (*10*), led to yields of about 39%, but the product also contained quantities of tri- and penta- chloro-substituted dibenzodioxins. Alternatively, the preparation and chlorination of 2,7-dichloro-*p*-dioxin (DCDD) (Figure 2), even though chlorination produces a less desirable product mix (*11, 12*), permitted the use of isotopic 2,4-dichlorophenol, an available starting material which was economically more expedient than 2,4,5-trichlorophenol-$^{14}C$. Experience with chlorinating dibenzo-*p*-dioxin while preparing analytical grade TCDD was helpful and provided confidence in this method which was ultimately used for the $^{14}C$ work.

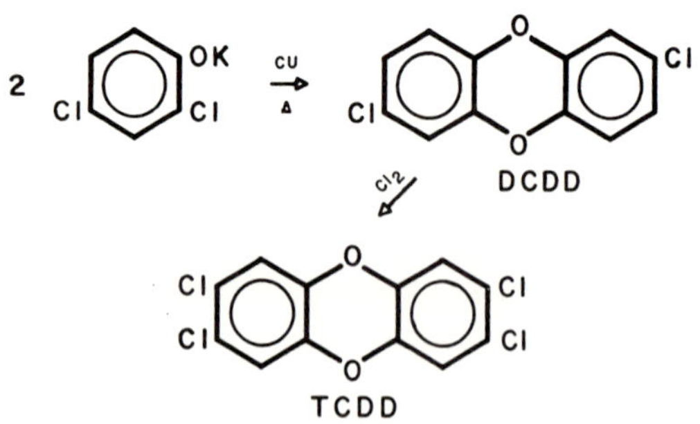


*Figure 2. Preparation of 2,3,7,8-tetrachlorodibenzo-*p*-dioxin from potassium 2,4-dichlorophenate*

### *Results and Discussion*

In pilot experiments DCDD was prepared by heating a mixture of potassium 2,4-dichlorophenate and a copper catalyst, prepared according to Brewster and Groening (*13*), in a stirred slurry with Decalin at 185°–210°C for 26 hours. The solid products, which separated after cooling the crude reaction mixture, were collected by filtration and washed with aqueous sodium hydroxide and with alcohol. A single recrystallization from anisole produced DCDD with an assay of 99.3%, as determined by mass spectroscopy. Response to ions with masses corresponding to mono- and tri-substituted dibenzodioxins were also detected in this analysis. Single crystal and X-ray powder diffraction data were taken (*14*) to confirm the structure. Yields approaching 20% were consistent with results reported by Julia and Baillarge (*15*) and by Villars (*16*). By-products consisted primarily of caustic soluble polymeric material with phenyl ether character; this suggests indiscriminate nucleophilic attack at the para position. Aniline (*10*) found that when potassium 2-bromo-4-chlorophenate was used in this reaction the by-product formation was reduced and the yields of DCDD increased to 70%. He attributed this to the more favorable kinetics offered by the ortho-bromo substituent, an observation reported by Weingarten (*17*) in a similar study.

**Table I. Composition of Products during the Chlorination of Dibenzo-*p*-dioxin**

| $C_{12}H_{8-n}Cl_nO_2$ | *Isomer Distribution (%) after:* | | | |
|---|---|---|---|---|
| n = | *2 Hours* | *5 Hours* | *17 Hours* | *24 Hours* |
| 1 | 30 | 5 | | |
| 2 | 61 | 32 | <1 | |
| 3 | 6 | 54 | 10 | |
| 4 | 2 | 7 | 88 | 54 |
| 5 | <1 | <1 | <1 | 43 |
| 6 | | | | 2 |

Uniformly labeled 2,4-dichlorophenol-$^{14}C$ (purchased from New England Nuclear Corp., Boston, Mass.) was used in the tracer preparation. This provided a label at all carbon positions in the dibenzodioxin structure. 2,7-Dichlorodibenzo-*p*-dioxin-$^{14}C$ after initial cleanup by fractional sublimation, contained approximately 5% of an impurity, detected by thin layer chromatography (TLC) which gave mass peaks at 288, 290, 292, and 294 in the mass spectrometer, consistent with a trichloro-hydroxydiphenyl oxide. This is probably the initial condensation product of the Ullman reaction and is most likely 2-(2,4-dichlorophenoxy)-4-chlorophenol. It was removed easily by extractions with aqueous sodium hydroxide solution from the impure product dissolved in chloroform. No other impurities were detected.

Experiments were conducted to determine reaction conditions for optimizing the yield of TCDD by chlorination. A 5% solution of dibenzo-*p*-dioxin in chloroform, containing catalytic amounts of $FeCl_3$ and $I_2$ and maintained at reflux temperature, was treated with chlorine gas for 24 hours. The flow was regulated to maintain a saturated solution. The composition of unrefined product with respect to time (Table I) was determined by mass spectroscopy. From these results chlorination seems to proceed rapidly to the tri-substituted species, less rapidly to the tetra-chloro product, and still less rapidly to a higher degree of substitution. Varying solubilities of the chlorinated products in chloroform probably contributed to this product distribution, as the higher chlorinated species are precipitated during the chlorination.

**Table II. Composition of Products during the Purification of Chlorinated Dibenzo-*p*-dioxin**

| $C_{12}H_{8-n}Cl_nO_2$ | *Isomer Distribution (%) after:* | | | | |
|---|---|---|---|---|---|
| n = | *0 Cycle* | *1 Cycle* | *2 Cycles* | *3 Cycles* | *4 Cycles* |
| 3 | 1 | 6.7 | 3.8 | 1.9 | 1.0 |
| 4 | 54 | 92 | 95 | 97.2 | 98.6 |
| 5 | 43 | 1.6 | 1.3 | 0.9 | 0.4 |
| 6 | 1.6 | | | | |

Advantage was taken of these solubility differences in refining mixtures of the chlorinated dibenzodioxins. Digestion with boiling chloroform was effective in removing trichlorodibenzodioxin while recrystallization from anisole reduced the penta-substituted isomer content. In a typical purification (Table II) these two procedures were alternated through four cycles. The assays were made by mass spectroscopy using the batch injection method to introduce the sample into the spectrometer. X-ray studies (*14*) confirmed the structure.

**Table III. Assay of 2,3,7,8-Tetrachlorodibenzo-*p*-dioxin-$^{14}C$ by Mass Spectroscopy**

| $C_{12}H_{8-n}Cl_nO_2$ | *Isomer Distribution (%) after:* | | | |
|---|---|---|---|---|
| n = | *Crude* | *Sublimation* | *Recrystal-lization* | *Extraction* |
| 3 | 19 | 13 | 11.6 | 5.2 |
| 4 | 79 | 85.5 | 86.9 | 93.3 |
| 5 | 2.1 | 1.7 | 1.6 | 1.5 |

Purification of the radioactive tracer was modified to include a fractional sublimation before a single extraction–recrystallization cycle to conserve the tracer material. Microgram samples were prepared in melting point capillaries for assay by mass spectroscopic analysis (Table III), made by direct probe injection of the sample into the ion source (*18*). The probe was heated rapidly to 200°C, and mass spectra were obtained during vaporization of the sample. Tri-, tetra-, and pentachlorodibenzo-*p*-dioxins vaporized simultaneously with no observed fractionation.

## *Experimental*

**2,7-Dichlorodibenzo-*p*-dioxin-$^{14}$C, U.L.** To a benzene solution of 4.89 grams of 2,4-dichlorophenol-$^{14}$C, U.L., containing 15.2 mCi of radioactivity, were added 10 ml of Decalin and 1.68 grams (1 equivalent) of potassium hydroxide dissolved in 10 ml of methanol. From this mixture, 20 ml of distillate were collected by heating to 150°C over 1 hour. After cooling, 0.5 gram of freshly prepared copper catalyst and 10 ml of benzene were added. Heating was resumed at 150°C for 30 min while an additional 10 ml of distillate were collected. The temperature was then raised to 185°C where it was maintained for 3 hours, then to 210°C for 12 hours. After cooling, the solid product was collected on a filter and washed with *n*-pentane. The product remaining on the filter was extracted seven times with 10-ml portions of hot choloroform. The extracts were combined in a separatory funnel where they were extracted with 25 ml of *N* NaOH. The lower organic layer was transferred to a sublimation flask, and the solvent was removed by distillation. The residue was sublimed at 120°C, providing 0.780 grams of impure DCDD. This was dissolved in 100 ml of fresh chloroform and extracted three times with 25 ml portions of *N* NaOH. Upon concentration of the chloroform layer to 20 ml and cooling, 0.165 gram of pure DCDD separated in long white needles. An additional 0.571 gram was obtained by further concentration of the mother liquor. The products contained 3.6 $\mu$Ci of radioactivity per mg.

**2,3,7,8-Tetrachlorodibenzo-*p*-dioxin-$^{14}$C, U.L.** 2,7-Dichlorodibenzo-*p*-dioxin-$^{14}$C, U.L. (0.500 gram) was stirred with 10 ml of chloroform containing trace amounts of $FeCl_3$ and $I_2$ and heated to reflux temperature while chlorine gas was passed into the mixture for 18 hours. After cooling, the white insoluble product was collected by filtration and triturated with 15 ml of boiling chloroform. The insoluble portion was transferred to a sublimation flask where it was vacuum-sublimed at 140°C. The sublimate was recrystallized from 2.5 ml of anisole and washed with chloroform. The product weighed 0.229 gram and contained 2.9 $\mu$Ci of radioactivity per mg.

**Mass Spectrometry.** The mass spectra were obtained on a CEC 21-110B mass spectrometer with the batch inlet system maintained at 250°C to assure complete vaporization of the samples. Sensitivity factors for quantitative analysis were obtained from standards of di-, tetra-, hexa-, and octa-chlorodibenzo-*p*-dioxin. The factors for the intermediate chlorinated species were estimated by interpolation. The analyses were based on peak heights for observed molecular ions (the most intense ion in the mass spectra of each dioxin analyzed).

The direct insertion probe consists of a metal sample holder drilled to accept standard melting point capillaries up to 1 inch in length. This is inserted into the ion source through a vacuum lock and may be heated to 250°C at varying rates.

**TLC Analysis.** Samples were examined by TLC using standard procedures. $R_f$ values were determined and compared with those of authentic reference materials. Radioactive components were located by scanning (Vanguard Instrument Corp., North Haven, Conn., Model 885) or by autoradiography (Eastman Kodak, Rochester, N. Y., type AA film). The relative $R_f$ value of DCDD on silica gel plates (Brinkmann Instruments, Inc., Westbury, N. Y., type $F_{254}$) when developed with *n*-hexane: dioxane:acetic acid, 90:10:4, V/V/V, was 0.90. The observed impurity had a relative $R_f$ value of 0.40. On Brinkmann alumina plates, developed with *n*-hexane, DCDD had a relative $R_f$ of 0.32. Neither system separated the chlorinated dibenzodioxin isomers.

### *Literature Cited*

1. Kimmig, J., Schulz, J. H., *Dermatologica* (1957) **115**, 540.
2. Jones, E. L., Krizek, H., *Invest. Dermatol.* (1962) **39**, 511.
3. Rowe, V. K., *et al.*, ADVAN. CHEM. SER. (1973) **120**, 55.
4. *Chem. Eng. News* (April 27, 1970) 60.
5. Woolson, E. A., Thomas, R. F., *161st Nat. Amer. Chem. Soc. Meet.*, Los Angeles (March, 1971).
6. Piper, W. N., Rose, J. Q., ADVAN. CHEM. SER. (1973) **120**, 85.
7. Kearney, P. C., *et al.*, ADVAN. CHEM. SER. (1973) **120**, 105.
8. Surrey, A. R., “Name Reactions in Organic Chemistry,” 2nd ed., p. 236, Academic, New York, 1961.
9. Langer, H., *et al.*, ADVAN. CHEM. SER. (1973) **120**, 26.
10. Aniline, O., personal communication, The Dow Chemical Co., Midland, Michigan (1970).
11. Sanderman, W., *et al.*, *Chem. Ber.* (1957) **90**, 690.
12. Tomita, M., *et al.*, *Yakaguka Zasski* (1959) **79**, 186.
13. Brewster, R. Q., Groening, T., “Organic Synthesis Collective Volume II,” p. 445, Wiley, New York, 1963.
14. Boer, F. P., *et al.*, ADVAN. CHEM. SER. (1973) **120**, 14.
15. Julia, M., Baillarge, M., *Bull. Soc. Chim. France* (1953) 640.
16. Villars, C. E., unpublished work, The Dow Chemical Co., Midland, Michigan (1955).
17. Weingarten, H., *J. Org. Chem.* (1964) **29**, 977.
18. Woolson, E. A., Thomas, R. F., *161st Nat. Amer. Chem. Soc. Meet.*, Los Angeles (March, 1971).

RECEIVED February 8, 1972.

# 2

# Preparation of Tritium-Labeled Dibenzo-*p*-dioxin and 2,3,7,8-Tetrachlorodibenzo-*p*-dioxin

J. HOWARD VINOPAL,[1] IZURU YAMAMOTO,[2] and JOHN E. CASIDA

Division of Entomology, University of California, Berkeley, Calif. 94720

*1,6-Ditritiodibenzo-*p*-dioxin, prepared by reductive dechlorination of 1,6-dichlorodibenzo-*p*-dioxin with tritium gas, is chlorinated in chloroform solution, containing catalytic amounts of iodine and ferric chloride, to produce 1,6-ditritio-2,3,7,8-tetrachlorodibenzo-*p*-dioxin.*

The highly toxic and teratogenic 2,3,7,8-tetrachlorodibenzo-*p*-dioxin (2,3,7,8-$Cl_4$-DBpD) may occur in trace amounts in the herbicide, 2,4,5-trichlorophenoxyacetic acid (2,4,5-T) (*1, 2*). Radiolabeled preparations of this dioxin are needed to facilitate studies of its degradation chemistry, metabolism, and mode of action.

Muelder and Shadoff (*3*) prepared $^{14}C$-2,3,7,8-$Cl_4$-DBpD (0.9 mCi/mmole) by chlorination of $^{14}C$-2,7-dichlorodibenzo-*p*-dioxin made from potassium $^{14}C$-2,4-dichlorophenate. The preparation of tritium-labeled 2,3,7,8-$Cl_4$-DBpD is justified because the radiolabeled intermediates are less expensive and more accessible and because a higher specific activity is potentially attainable. Here, we consider the optimal conditions for the reaction sequence designed to obtain products of high chemical and radiochemical purity shown at the top of p. 8.

### *Experimental*

**Designation, Chromatographic Separation, and Analysis.** The abbreviation DBpD is used for dibenzo-*p*-dioxin, and the positions of chlorine

[1] Present address: Entomological Sciences and Pesticide Division, U.S. Army Environmental Hygiene Agency, Edgewood Arsenal, Md. 21010.

[2] Present address: Department of Agricultural Chemistry, Tokyo University of Agriculture, Setagaya-ku, Tokyo, Japan.

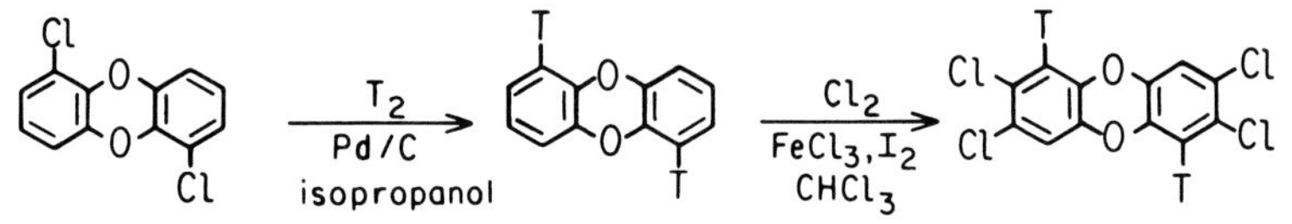


1,6-$Cl_2$-DBpD 1,6-$T_2$-DBpD 1,6-$T_2$-2,3,7,8-$Cl_4$-DBpD

(Cl) or tritium (T) in its derivatives are indicated by appropriate designations—*e.g.*, 1,6-$Cl_2$-DBpD, 1,6-$T_2$-DBpD, and 1,6-$T_2$-2,3,7,8-$Cl_4$-DBpD.

The composition and purity of the unlabeled reaction products were determined by gas–liquid chromatography (GLC), using a flame ionization detector under the conditions given in Table I. Radio–GLC analysis of the labeled products was made by the Zoecon Corp. (Palo Alto, Calif.), using a Barber–Coleman instrument (model Selecta System, series 5000; Rockford, Ill.) and a glass column (1 meter × 3 mm) containing SE-30 (3.8%, w/w) on Chromosorb W (80/100 mesh). A set of conditions (injection and column temperature, 110°C; programmed for 3°C increase in temperature/minute; $He_2$, 60 ml/min), which separate DBpD, 1-Cl-DBpD, and 1,6-$Cl_2$-DBpD and another set of conditions (injection and column temperature, 220°C; $He_2$, 60 ml/min), which separate $Cl_3$-DBpD, 2,3,7,8-$Cl_4$-DBpD, and $Cl_5$-DBpD, were used for radio–GLC. Thin layer chromatography (TLC) on Eastman 6061 silica gel chromagram sheets (0.1 mm gel thickness) is useful in separating the DBpD derivatives (high $R_f$) from the corresponding phenols (low $R_f$) but not from each other because the $R_f$ values are the same for the DBpD derivatives (DBpD, 1-Cl-DBpD, 1,6-$Cl_2$-DBpD, 2,7-$Cl_2$-DBpD, 2,3,7,8-$Cl_4$-DBpD). The $R_f$ values found with each of these derivatives are 0.54–0.55 on development with carbon tetrachloride, 0.69–0.71 with hexane, 0.81–0.82 with acetone, 0.95 with benzene, and 1.0 with ether. Unlabeled compounds were detected with a silver nitrate–hydrogen peroxide reagent (*4*) (sensitive to 0.5 $\mu$g DBpD and 0.1 $\mu$g chloro-DBpD derivative), and labeled compounds were traced by radioautography.

Mass spectra (MS) were determined with direct inlet insertion of the samples in a Type 21-103C mass spectrometer (Consolidated Electrodynamics Corp., Pasadena, Calif.). In the analysis of samples containing a mixture of $Cl_3$-DBpD, 2,3,7,8-$Cl_4$-DBpD, and $Cl_5$-DBpD, the ratios of mass peak 285 (owing to $Cl_3$-DBpD but not 2,3,7,8-$Cl_4$-DBpD) and mass peak 358 (owing to $Cl_5$-DBpD) to mass peak 322 (from 2,3,7,8-$Cl_4$-DBpD) were determined and corrected for the sensitivity factors of the respective parent ions of these compounds. The appropriate sensitivity factors were provided by the Dow Chemical Co. (Midland, Mich.) (*5*). The crystal structure studies, involving x-ray diffraction (*6*), were performed by the Dow Chemical Co.

The chemical purity and identity of the unlabeled and labeled chemicals were determined by infrared spectroscopy, mass spectrometry, GLC, and TLC.

**Unlabeled Chemicals.** Samples of unlabeled DBpD derivatives were obtained from the following sources: 2,3,7,8-$Cl_4$-DBpD (analytical standard, >99% pure) from the Dow Chemical Co.; 1-Cl-DBpD and 2,3,7,8-$Cl_4$-DBpD from A. E. Pohland (Division of Chemistry and Physics, U. S. Food and Drug Administration, Washington, D. C.).

DBpD was prepared by slowly heating a mixture of *o*-chlorophenol (480 mmoles), potassium carbonate (240 mmoles), and purified copper powder (50 mmoles) in a 500-ml Erlenmeyer flask to 160°–180°C and maintaining this temperature for 6 hours. DBpD, which sublimed to the walls of the flask as it was formed, was recovered by scraping and was recrystallized from absolute ethanol (14% yield; white needles, mp 119°–120°C; reported 120°–122°C (7); elemental analyses—calcd. C = 78.26, H = 4.34, found C = 77.98, H = 4.48).

DBpD was also prepared on a submillimole-scale by reductive dechlorination of 1,6-$Cl_2$-DBpD. In this method 1,6-$Cl_2$-DBpD (0.1 mmole) in 2-propanol (5 ml) containing 10% palladium on powdered charcoal (25 mg) was hydrogenated at a pressure of 10 lbs/square inch with shaking for 1 hour at 25°C in the 200-ml pressure bottle of a conventional hydrogenation apparatus (Parr Instrument Co., Moline, Ill.). The catalyst was removed by filtration and was washed with 25 ml of 2-propanol. The combined 2-propanol fractions were evaporated onto Florisil (10 gram, 60/100 mesh, Floridin Co., Berkeley Springs, W. Va.), and the mixture was added to a glass column (4.5 cm diameter) containing 30 grams Florisil. Then 25 grams Florisil were added to the top of the column, and DBpD was eluted with hexane which left the very polar (probably phenolic) breakdown products on the column. The appropriate fractions were combined (based on TLC- and GLC-monitoring) and evaporated under nitrogen to give pure DBpD (55–65% yield; single compound by GLC and TLC; mp 119°C, without recrystallization).

1,6-$Cl_2$-DBpD was prepared either by heating 2,6-dichlorophenol (150 mmoles), potassium carbonate, and copper powder under the conditions above for DBpD or by heating the potassium phenate (35 mmoles)

**Table I. GLC Characteristics of Dibenzo-*p*-dioxin and its Chlorinated Derivatives**

| Compound | Retention Time, min | |
|---|---|---|
| | *1% SE-30* [a] | *10% OV-17* [b] |
| DBpD | 1.5 [c], — | 3.0 [e], 0.8 [f] |
| 1-Cl-DBpD | 3.0 [c], — | 6.0 [e], 1.4 [f] |
| 1,6-$Cl_2$-DBpD | 5.9 [c], 0.7 [d] | 11.0 [e], 2.4 [f] |
| 2,7-$Cl_2$-DBpD | 5.9 [c], 0.7 [d] | 10.0 [e], 2.2 [f] |
| 2,3,7-$Cl_3$-DBpD [g] | — , 1.3 [d] | — , — |
| 2,3,7,8-$Cl_4$-DBpD | — , 2.5 [d] | 33.5 [e], 6.8 [f] |

[a] On Chromosorb P (100/120 mesh), 1.5 m × 3 mm stainless steel column
[b] On Chromosorb W (80/100 mesh), 1.5 m × 3 mm stainless steel column
[c] Injection temperature = 245°C, column temperature = 160°C, $N_2$ = 25 ml/min
[d] Injection temperature = 280°C, column temperature = 195°C, $N_2$ = 66 ml/min
[e] Injection temperature = 280°C, column temperature = 250°C, $N_2$ = 25 ml/min
[f] Injection temperature = 280°C, column temperature = 270°C, $N_2$ = 66 ml/min
[g] Obtained by irradiating an ethanol solution of 2,3,7,8-$Cl_4$-DBpD with short-wavelength ultraviolet light.

directly with copper powder. The desired product sublimed in the former method at 240°–260°C and in the latter one at 300°–320°C. After recrystallization from methanol, the products obtained had the following characteristics: by the phenol method—4.1% yield; white plates, mp 197°–199°C; elemental analyses—calcd. C = 56.91, H = 2.37, Cl = 28.06, found C = 57.04, H = 2.52, Cl = 28.18, and by the potassium phenate procedure—7.8% yield; white needles, mp 198°–200°C; elemental analyses—found C = 56.92, H = 2.30, Cl = 27.88.

2,7-$Cl_2$-DBpD was prepared by the same two procedures used for 1,6-$Cl_2$-DBpD involving 2,4-dichlorophenol (75 mmoles) and sublimation at 200°–220°C or potassium 2,4-dichlorophenate (63 mmoles) and sublimation at 220°–240°C. After recrystallization from petroleum ether, the products obtained had the following characteristics: by the phenol procedure—4.5% yield; white needles, mp 206°–208°C; elemental analyses—calcd. C = 56.91, H = 2.37, Cl = 28.06, found C = 56.64, H = 2.40, Cl = 28.10, and by the potassium phenate procedure—5.7% yield; white needles, mp 207°–208°C; elemental analyses—found C = 56.92, H = 2.52, Cl = 28.12 (2,7-$Cl_2$-DBpD prepared previously from sodium 2,4-dichlorophenate gave mp 201°–202°C (8) and by chlorination of DBpD gave mp 203°C (7) and mp 195°–197°C (9)).

2,3,7,8-$Cl_4$-DBpD was prepared on a submillimole-scale by chlorination of DBpD under carefully standardized conditions. Trace amounts of iodine and ferric chloride were added to a solution of pure DBpD (0.22–0.24 mmole) in chloroform (0.5 ml) in a 25-ml conical test tube. After cooling to 0°–5°C, chlorine gas was bubbled slowly through the chloroform mixture for 5–6 min, forming a copious precipitate. The precipitate in the reaction mixture was washed twice with 1.5-ml portions of chloroform at 25°C, using centrifugation for maximum separation, to remove the unwanted chloroform-soluble materials. The precipitate was then treated by shaking it with chloroform (5 ml) and water (1 ml), and the water-soluble products (such as ferric chloride) were removed, following centrifugation. The chloroform phase, containing some undissolved material, was evaporated to dryness under a nitrogen stream, and the residue was transferred to a 25-ml Erlenmeyer flask to which was added anisole (10 ml). The anisole in the flask was heated to reflux to dissolve most of the residue, and the undissolved portion was removed by filtering the hot anisole solution. Crystallization was achieved by allowing the anisole solution to cool to 25°C, putting it in the refrigerator overnight, and filtering to recover 0.0093–0.0185 mmole of material which contained (GLC and MS) 90–94% 2,3,7,8-$Cl_4$-DBpD (range from several different preparations), the remainder being mostly $Cl_3$- and $Cl_5$-DBpD. In one study three batches of material (0.050 mmole) from the anisole recrystallization were combined and washed with hot chloroform (8 ml). The chloroform-insoluble portion (0.022 mmole) consisted of a single crystalline material which gave a x-ray diffraction pattern identical to that of known 2,3,7,8-$Cl_4$-DBpD, indicating that the product consisted only of 2,3,7,8-$Cl_4$-DBpD and did not contain any other isomeric $Cl_4$-DBpD. (The product also contained some glass, possibly from the coarse fritted disk in the Buchner funnel used in filtrations.) In another study the product from the anisole recrystallization (0.013 mmole) was combined with authentic 2,3,7,8-$Cl_4$-DBpD (0.037 mmole, >99% pure) to obtain sufficient material to wash twice with hot chloroform (8 ml and 3 ml). The final product was >99% 2,3,7,8-$Cl_4$-DBpD, based on MS analysis.

**Labeled Chemicals.** 1,6-$T_2$-DBpD was synthesized essentially by the same procedure used to prepare unlabeled DBpD on a submillimole scale by reductive dechlorination (above) with the following exceptions. After reduction of 1,6-$Cl_2$-DBpD (0.1 mmole) in 2-propanol (5 ml) for 1 hour by stirring with tritium gas (5 Ci) and hydrogen gas (0.5 ml) at atmospheric pressure, the reaction mixture was washed with 2-propanol (2 × 10 ml) to remove labile tritium; the catalyst was removed by filtration; the solvent was removed by vacuum distillation, and the products were redissolved in 2-propanol (10 ml). (This portion of the procedure was carried out by New England Nuclear Corp. (Boston, Mass.).) The tritiated material was diluted (to obtain an appropriate scale of reaction in subsequent steps) by adding pure unlabeled DBpD (0.082 mmole) in 2-propanol (25 ml), and the resulting mixture was chromatographed (as above) to isolate crude 1,6-$T_2$-DBpD (0.17 mmole, 1.1 Ci/mmole). Analyses by MS and GLC indicated the presence of trace amounts of a contaminant, 1-Cl-6-T-DBpD, in the crude 1,6-$T_2$-DBpD. The presence of this kind of contaminant in the labeled preparation but not in the unlabeled preparation resulted probably from the lower pressure used during the reduction step—*i.e.* 10 lbs/square inch for the unlabeled preparation and atmospheric pressure for the tritium-preparation. A small portion (0.015 mmole) of the crude 1,6-$T_2$-DBpD was reduced to remove the trace level of monochloro impurity by dissolving the sample in 2-propanol (3.5 ml), adding 10% palladium on powdered charcoal (3.8 mg), and shaking with hydrogen gas at 8 lbs/square inch for 15 min at 25°C. The catalyst was removed by filtration, and an additional portion of pure unlabeled DBpD (0.054 mmole) in 2-propanol (10 ml) was added before chromatographic isolation of 1,6-$T_2$-DBpD (138 mCi/mmole; single radioactive compound by TLC and radio-GLC; no 1-Cl-DBpD detected by MS). The calculated specific activity, assuming that there is no loss of compound on reduction, is 239 mCi/mmole, indicating that a portion of the labeled material is lost during the process of the second reduction.

1,6-$T_2$-2,3,7,8-$Cl_4$-DBpD was prepared by chlorination of 1,6-$T_2$-DBpD (0.23 mmole, a mixture of 0.15 mmole of the crude 1,6-$T_2$-DBpD described above and 0.08 mmole of unlabeled DBpD) in the same manner described previously for preparing unlabeled 2,3,7,8-$Cl_4$-DBpD. This resulted in the isolation of crude 1,6-$T_2$-2,3,7,8-$Cl_4$-DBpD (0.0115 mmole, calculated as the tetrachloro derivative although some $Cl_3$-DBpD and a smaller amount of $Cl_5$-DBpD were present) following recrystallization from anisole. Unlabeled 2,3,7,8-$Cl_4$-DBpD (0.043 mmole) was then added to the crude labeled product, and the mixture was washed with hot chloroform (8 ml), as described above for the unlabeled 2,3,7,8-$Cl_4$-DBpD, filtered, and washed again with hot chloroform (3 ml) to obtain the final 1,6-$T_2$-2,3,7,8-$Cl_4$-DBpD preparation (0.011 mmole, 107 mCi/mmole). The specific activity is below the theoretical value of 152 mCi/mmole, calculated from the various dilution factors, probably as a result of inaccuracies in weighings or of a small amount of glass in the sample from the fritted glass filter. The product contained no radiolabeled impurities detectable by TLC and radioautography or radio–GLC; there were no significant unlabeled contaminants evident on GLC, and MS indicated a chemical purity of >99%.

### *Discussion*

Three routes for preparing tritium-labeled 2,3,7,8-$Cl_4$-DBpD were investigated. The first involved an exchange reaction between tritium water and 2,4,5-trichlorophenol to obtain tritium-labeled 2,4,5-trichlorophenol, useful in synthesizing 2,3,7,8-$Cl_4$-DBpD by the salt fusion reaction. Preliminary studies with deuterium oxide, involving proton magnetic resonance monitoring, indicated that extensive exchange of aromatic protons can be achieved under basic but not acidic conditions. However, the degree of exchange does not seem to be sufficiently great to prepare high specific activity material upon replacing the deuterium oxide with tritium water. The second route started with 1,6-$Cl_2$-DBpD which was converted to the 1,6-dilithium derivative, and this was hydrolyzed with tritium water to produce 1,6-$T_2$-DBpD. The DBpD prepared by this method was always contaminated with about 30% 1-Cl-DBpD, indicating that the overall reaction was not sufficiently complete. Also, the separation of DBpD and 1-Cl-DBpD was very difficult. The route selected uses tritium gas instead of tritium water for labeling, and this change allows compounds of higher specific activity to be prepared.

Reductive dechlorination of 1,6-$Cl_2$-DBpD, using tritium gas and subsequent chlorination of the 1,6-$T_2$-DBpD, is a convenient method for preparing 1,6-$T_2$-2,3,7,8-$Cl_4$-DBpD, but care must be exercised in several steps involved in this submillimole-scale radiosynthesis. Insufficient reduction of 1,6-$Cl_2$-DBpD results in contamination by 1-Cl-DBpD while the amount of polar products increases if the reduction is too drastic. Using exactly 0.5 ml of precooled chloroform for the chlorination step results in a mixture of the following components, as determined by GLC chromatography and MS: DBpD; a monochloro derivative (probably 2-Cl-DBpD); 2,7-$Cl_2$-DBpD; another isomeric $Cl_2$-DBpD; a trichloro derivative, assumed to be 2,3,7-$Cl_3$-DBpD; 2,3,7,8-$Cl_4$-DBpD, and $Cl_5$-DBpD. No higher chlorinated derivatives of DBpD were detected. The contaminants in the mixture are removed largely in various purification stages. Washing the crude reaction mixture with chloroform removes all unreacted DBpD, most of the 1-Cl-DBpD, and some of the $Cl_2$- and $Cl_3$-DBpD derivatives. The next step of washing the chloroform with water removes the ferric chloride catalyst. Following anisole recrystallization of the chloroform-soluble products, the material consists of 90–94% of 2,3,7,8-$Cl_4$-DBpD and 6–10% of $Cl_3$- and $Cl_5$-DBpD. The final chloroform washes remove almost all residual $Cl_3$- and $Cl_5$-DBpD, resulting in very pure 2,3,7,8-$Cl_4$-DBpD.

The samples of 1,6-$T_2$-DBpD and 1,6-$T_2$-2,3,7,8-$Cl_4$-DBpD are useful in metabolism and mode of action studies. For example, when incubated with rabbit liver microsomes, 1,6-$T_2$-DBpD is extensively metabolized to polar product(s) but only when these preparations are fortified with reduced nicotinamide–adenine dinucleotide phosphate. Under the same conditions 1,6-$T_2$-2,3,7,8-$Cl_4$-DBpD is completely resistant to metabolic attack. In some types of studies, a higher specific activity possibly is desirable (*i.e.*, >1 Ci/mmole), and this can be achieved, with the methodology already developed, by using larger amounts of tritium gas or working on a larger synthetic scale so that it is not necessary to add unlabeled materials to assist in crystallization steps where a certain minimum amount of compound is necessary.

### *Acknowledgment*

This work was aided by PHS grant ES 00049 and AEC Contract No. AT(04-3)-34, project agreement No. 113. The authors are indebted to June Turley and Lewis Shadoff (Chemical Physics Research Laboratory) and Warren Crummett (Analytical Laboratory, The Dow Chemical Co., Midland, Mich.) and to Loren Dunham (Zoecon Corp., Palo Alto, Calif.) for advice and assistance in product analysis.

### *Literature Cited*

1. Wilson, J. G., *et al.*, "Report of the Advisory Committee on 2,4,5-T to the Administrator of the Environmental Protection Agency" (May 7, 1971).
2. Schwetz, B. A., Norris, J. M., Sparschu, G. L., Rowe, V. K., Gehring, P. J., Emerson, J. L., Gerbig, C. G., ADVAN. CHEM. SER. (1973) **120,** 55.
3. Muelder, W. W., Shadoff, L. A., ADVAN. CHEM. SER. (1973) **120,** 1.
4. Kovacs, M. F., Jr., *J. Ass. Offic. Agr. Chem.* (1963) **46,** 884.
5. Dow Chemical Co., private communication.
6. Boer, F. P., Neuman, M. A., van Remoortere, F. P., North, P. P., Rinn, H. W., ADVAN. CHEM. SER. (1973) **120,** 14.
7. Gilman, H., Dietrich, J. J., *J. Amer. Chem. Soc.* (1957) **79,** 1439.
8. Julia, M., Baillarge, M., *Bull. Soc. Chim. France* (1953) 644.
9. Tomita, M., Ueda, S., Narisada, M., *Yakugaku Zasshi* (1959) **79,** 186.

RECEIVED February 18, 1972.

# 3

# X-ray Diffraction Studies of Chlorinated Dibenzo-*p*-dioxins

F. P. BOER, M. A. NEUMAN, F. P. VAN REMOORTERE, P. P. NORTH, and H. W. RINN

Analytical Laboratories, The Dow Chemical Co., Midland, Mich. 48640

*The crystal structures of four chlorinated derivatives of dibenzo-p-dioxin have been determined by x-ray diffraction from diffractometer data (MoKα radiation). The compounds, their formulae, cell dimensions, space groups, the number of molecules per unit cell, the crystallographic* R*-factors, and the number of observed reflections are given. The dioxin crystal structures were performed to provide absolute standards for assignment of isomeric structures and have been of considerable practical use in combination with x-ray powder diffraction analysis.*

Because of the present interest in the chemical, biological, and ecological properties of the chlorinated dibenzo-*p*-dioxins, we have applied X-ray diffraction methods to identify and characterize these compounds. Such studies are useful for three reasons:

(1) Absolute identification of the isomeric structure of primary standard materials by single crystal techniques;

(2) Correlation of the single-crystal structure results with the powder diffraction pattern, establishing x-ray powder diffraction as a convenient, powerful means of isomer identification; and

(3) Generation of a (usually) accurate, detailed picture of the molecular geometry to permit future correlations between structural features and biological activity.

We report the crystal structures of four chlorinated dioxins—the 2,7-dichloro-, 2,8-dichloro-, 2,3,7,8-tetrachloro-, and octachlorodibenzo-*p*-dioxins. Thus, five crystal structures of chlorodioxins are now known.

The unexpected identification (*1*) of 1,2,3,7,8,9-hexachlorodibenzo-*p*-dioxin in toxic feed fats by Cantrell, Webb, and Mabis, using x-ray crystallography, first pointed to the chlorodioxins as a potential toxicological hazard. The four crystal structures reported here were determined by very similar techniques, and their essential parameters are summarized in Table I. A Picker four-circle diffractometer was used to determine the lattice constants and to gather intensity data in $\theta$–$2\theta$ scan mode, using monochromatic Mo$K\alpha$ radiation. The structures were solved from Patterson functions and refined by full-matrix least-squares, assuming anisotropic temperature factors for Cl, O, and C atoms and isotropic temperature factors for H. Hydrogen parameters were not refined for the 2,8-dichloro compound for which only a small fraction of the possible reflections was obtained owing to the weakness of scattering and progressive deterioration of the crystal. Corrections were applied for absorption, for anomalous scattering by Cl, and, where appropriate, for secondary extinction. Table II lists the atomic positions and thermal parameters for the four structures.

The molecules of the 2,7-dichloro, 2,3,7,8-tetrachloro, and octachloro structures are all located on crystallographic centers of symmetry and are very nearly planar in each case. (In the 2,3,7,8-tetrachloro structure there are two independent molecules situated on the $\bar{1}$ elements at the origin and at the center of the unit cell, respectively.) In each of these three structures the molecules form stacks along the short unit cell translations with interplanar distances between 3.45–3.49A. The structure of the 2,8-dichloro compound is a different type with the molecule divided transversely by a crystallographic mirror plane and exhibiting a slight bending at the oxygens, similar to that described for the 1,2,3,7,8,9-hexa compound (*1*). Figures 1, 2, 3, and 4 indicate that the molecular geometries are quite regular with bond distances and angles agreeing with accepted literature values. The C—Cl distances, however, show an interesting trend toward shortening with increasing chlorine substitution on the rings (Figure 5). The estimated errors and internal agreement of the bond distances give confidence that this effect is real. The differences also seem to hold when bond distance corrections assuming rigid body motion are applied (*2*). This result could arise from a reduction in the effective electronegativity difference between C and Cl as more electron density is drawn from the aromatic ring, which should in turn result in increased covalency of the C—Cl bonds and give shorter distances. Alternatively, in terms of molecular orbitals, the addition of more electronegative chlorine substituents could draw electrons from antibonding MO's associated with the C—Cl bonds and thus increase the net bond order of the C—Cl bonds.

**Table I.**

| | | 2,7-DCBD |
|---|---|---|
| *Space Group* | | $P\bar{1}$ |
| Cell Constants | a (A) | 3.878(3) |
| | b (A) | 6.755(9) |
| | c (A) | 10.265(15) |
| | α (deg) | 99.46(1) |
| | β (deg) | 100.63(3) |
| | γ (deg) | 99.73(3) |
| | Z | 1 |
| Density | (g cm$^{-3}$) | 1.647 |
| Linear absorption coeff. (cm$^{-1}$) | | 5.26 |
| No. Reflections Measured | | 1152 |
| No. Reflections Above Background | | 1030 |
| $R_1 = \Sigma \mid \mid F_o \mid - \mid \mid F_c \mid \mid / \Sigma \mid F_o \mid$ | | 0.057 |
| $R_2 = [\Sigma w(F_o - F_c)^2 / \Sigma w F_o^2]^1/^2$ | | 0.062 |

**Table II. Structure**

*2,7-Dichlorodibenzo-p-dioxin*

| *Atom* | x | y | z |
|---|---|---|---|
| Cl | −0.4292(3) | −0.3654(1) | 0.3777(1) |
| O | 0.0466(7) | 0.2155(3) | 0.0479(2) |
| C(1) | −0.0668(8) | 0.0749(5) | 0.1228(3) |
| C(2) | −0.1125(8) | −0.1359(5) | 0.0757(3) |
| C(3) | −0.1340(9) | 0.1467(5) | 0.2468(3) |
| C(4) | −0.2219(9) | −0.2706(5) | 0.1534(3) |
| C(5) | −0.2452(9) | 0.0123(5) | 0.3255(3) |
| C(6) | −0.2894(8) | −0.1955(5) | 0.2782(3) |
| H(3) | −0.094(4) | 0.293(5) | 0.278(3) |
| H(4) | −0.240(9) | −0.411(6) | 0.124(4) |
| H(5) | −0.283(9) | 0.058(5) | 0.406(4) |

*2,8–Dichlorodibenzo-p-dioxin*

| *Atom* | x | y | z |
|---|---|---|---|
| Cl | 0.3003(6) | 0.0705(7) | 0.4541(1) |
| O(1) | 0.2586(20) | 0.0973(21) | 0.25 |
| O(2) | −0.1758(22) | −0.0780(22) | 0.25 |
| C(1) | 0.1492(25) | 0.0443(18) | 0.2969(6) |
| C(2) | −0.0638(22) | −0.0372(18) | 0.2971(5) |
| C(3) | 0.2607(25) | 0.0804(21) | 0.3449(6) |
| C(4) | −0.1596(20) | −0.0843(18) | 0.3460(6) |
| C(5) | 0.1541(31) | 0.0298(19) | 0.3937(5) |
| C(6) | −0.0581(27) | −0.0489(23) | 0.3944(6) |

## Crystal Data

| 2,8-DCBD | 2,3,7,8-TCBD | OCBD |
|---|---|---|
| *Pnam* | $P\bar{1}$ | $P2_1/c$ |
| 5.983(6) | 3.783(3) | 12.009(9) |
| 7.114(10) | 9.975(9) | 3.828(3) |
| 24.637(36) | 15.639(15) | 16.297(9) |
| 90. | 94.14(2) | 90. |
| 90. | 95.20(4) | 101.14(3) |
| 90. | 92.77(4) | 90. |
| 4 | 2 | 2 |
| 1.602 | 1.827 | 2.077 |
| 5.12 | 9.79 | 15.05 |
| 1206 | 2666 | 1640 |
| 293 | 2381 | 1460 |
| 0.064 | 0.036 | 0.058 |
| 0.065 | 0.038 | 0.053 |

## Parameters[a, b]

| $10^4\beta_{11}$ | $10^4\beta_{22}$ | $10^4\beta_{33}$ | $10^4\beta_{12}$ | $10^4\beta_{13}$ | $10^4\beta_{23}$ |
|---|---|---|---|---|---|
| 945(9) | 240(2) | 106(1) | 85(3) | 142(2) | 63(1) |
| 1110(25) | 134(5) | 72(2) | 82(8) | 114(6) | 15(3) |
| 587(26) | 159(7) | 66(3) | 74(11) | 36(7) | 30(4) |
| 569(26) | 165(2) | 66(3) | 77(11) | 35(7) | 21(4) |
| 694(29) | 166(8) | 78(4) | 74(12) | 64(8) | 3(4) |
| 599(28) | 160(9) | 84(4) | 63(11) | 41(8) | 32(4) |
| 640(29) | 213(9) | 77(4) | 110(12) | 84(8) | 25(5) |
| 453(24) | 208(8) | 89(4) | 89(11) | 55(7) | 58(4) |

| *B* |
|---|
| 3.5(8) |
| 4.4(9) |
| 4.0(9) |

| $10^3\beta_{11}$ | $10^3\beta_{22}$ | $10^4\beta_{33}$ | $10^3\beta_{12}$ | $10^3\beta_{13}$ | $10^3\beta_{23}$ |
|---|---|---|---|---|---|
| 29(1) | 37(1) | 15(1) | −4(1) | −1(1) | 0(1) |
| 17(6) | 32(5) | 10(2) | −10(4) | | |
| 15(5) | 28(5) | 18(3) | −7 | | |
| 20(6) | 10(3) | 13(4) | −3(3) | −1(1) | 0(1) |
| 9(6) | 17(4) | 17(4) | −4(4) | 2(2) | −1(1) |
| 26(6) | 13(3) | 14(3) | 0(4) | 1(1) | 1(1) |
| 10(5) | 9(3) | 20(3) | −3(4) | 1(1) | 0(1) |
| 30(7) | 16(4) | 10(2) | 10(5) | −1(1) | −1(1) |
| 8(5) | 26(4) | 15(3) | 4(5) | 1(1) | −1(1) |

**Table II.**

| *Atom* | x | y | z |
|---|---|---|---|
| H(3) | 0.414 | 0.141 | 0.345 |
| H(4) | −0.311 | −0.148 | 0.346 |
| H(6) | −0.136 | −0.080 | 0.430 |

*2,3,7,8-Tetrachlorodibenzo-*p*-dioxin*

| *Atom* | x | y | z |
|---|---|---|---|
| Cl(2A) | 0.24332(15) | −0.28365(5) | 0.26900(3) |
| Cl(3A) | −0.17456(16) | −0.47674(5) | 0.11985(4) |
| Cl(2B) | 0.10613(16) | 0.77903(6) | 0.77142(3) |
| Cl(3B) | 0.17045(17) | 0.97910(6) | 0.62508(4) |
| O(A) | 0.19947(40) | 0.08571(14) | 0.06607(9) |
| O(B) | 0.46780(42) | 0.40986(13) | 0.56599(9) |
| C(1A) | 0.20316(53) | −0.09398(20) | 0.15576(13) |
| C(2A) | 0.11584(51) | −0.22616(20) | 0.16989(12) |
| C(3A) | −0.06777(52) | —0.31053(19) | 0.10454(13) |
| C(4A) | −0.17010(53) | −0.26211(20) | 0.02588(13) |
| C(5A) | −0.08608(50) | −0.13010(19) | 0.01266(12) |
| C(6A) | 0.10264(50) | −0.04631(19) | 0.07732(12) |
| C(1B) | 0.29579(55) | 0.58979(21) | 0.65628(13) |
| C(2B) | 0.22673(51) | 0.72370(21) | 0.67145(13) |
| C(3B) | 0.25627(52) | 0.81109(20) | 0.60742(13) |
| C(4B) | 0.35698(56) | 0.76516(21) | 0.52812(13) |
| C(5B) | 0.42831(50) | 0.63162(20) | 0.51347(12) |
| C(6B) | 0.39708(52) | 0.54373(19) | 0.57761(12) |
| H(1A) | 0.3493(57) | −0.0369(23) | 0.1996(14) |
| H(4A) | −0.3102(52) | −0.3208(20) | −0.0219(13) |
| H(1B) | 0.2730(53) | 0.5287(21) | 0.7043(13) |
| H(4B) | 0.3803(53) | 0.8316(21) | 0.4815(13) |

*Octachlorodibenzo-*p*-dioxin*

| *Atom* | x | y | z |
|---|---|---|---|
| Cl(1) | 0.36887(8) | 0.1125(3) | 0.23425(6) |
| Cl(2) | 0.44176(8) | −0.2322(3) | 0.07846(7) |
| Cl(3) | 0.11407(8) | 0.3190(3) | 0.22562(6) |
| Cl(4) | 0.25867(8) | −0.3520(3) | −0.08448(6) |
| O | −0.0329(2) | 0.1580(10) | 0.0705(2) |
| C(1) | 0.0773(3) | 0.0630(12) | 0.0704(2) |
| C(2) | 0.1098(3) | −0.0887(12) | 0.0017(2) |
| C(3) | 0.1574(3) | 0.1261(11) | 0.1419(2) |
| C(4) | 0.2222(3) | −0.1749(11) | 0.0033(2) |
| C(5) | 0.2699(2) | 0.0359(11) | 0.1449(2) |
| C(6) | 0.3031(3) | −0.1153(11) | 0.0758(2) |

[a] The anisotropic thermal parameters are in the form $\exp\text{-}(h^2\beta_{11} + k^2\beta_{22} + l^2\beta_{33} + 2hk\beta_{33} + 2hl\beta_{13} + 2kl\beta_{23})$

**Continued**

| $10^3\beta_{11}$ | $10^3\beta_{22}$ | $10^4\beta_{33}$ | $10^3\beta_{12}$ | $10^3\beta_{13}$ | $10^3\beta_{23}$ |
|---|---|---|---|---|---|
| *B* | | | | | |
| 4.0 | | | | | |
| 4.0 | | | | | |
| 4.0 | | | | | |

| $10^4\beta_{11}$ | $10^5\beta_{22}$ | $10^5\beta_{33}$ | $10^5\beta_{12}$ | $10^5\beta_{13}$ | $10^5\beta_{23}$ |
|---|---|---|---|---|---|
| 758(5) | 865(6) | 301(2) | 325(13) | 130(8) | 132(3) |
| 841(5) | 617(5) | 407(2) | −179(13) | 152(9) | 78(3) |
| 803(5) | 924(6) | 310(2) | 455(14) | 463(8) | 95(3) |
| 864(5) | 635(5) | 469(3) | 645(13) | 340(10) | 48(3) |
| 800(13) | 629(15) | 248(6) | 386(35) | −217(22) | 16(7) |
| 920(14) | 620(15) | 244(6) | 682(36) | 359(23) | 56(7) |
| 465(14) | 678(21) | 256(8) | 220(43) | 9(27) | −25(10) |
| 464(14) | 723(21) | 255(8) | 435(43) | 98(27) | 76(10) |
| 472(14) | 545(19) | 325(9) | 108(42) | 231(28) | 36(10) |
| 463(14) | 635(20) | 276(9) | 3(42) | 79(27) | −43(10) |
| 437(14) | 656(20) | 225(8) | 116(41) | 87(25) | 12(9) |
| 453(14) | 587(19) | 253(8) | 119(41) | 126(26) | −6(10) |
| 529(16) | 731(22) | 256(8) | 168(46) | 148(28) | 8(11) |
| 446(14) | 762(22) | 261(8) | 161(44) | 122(27) | −74(10) |
| 452(15) | 594(20) | 346(9) | 265(42) | 5(29) | 44(11) |
| 556(16) | 648(21) | 305(9) | 226(45) | 113(29) | 51(11) |
| 446(14) | 665(20) | 236(8) | 272(41) | 68(26) | 5(10) |
| 473(14) | 604(20) | 267(8) | 242(41) | 35(27) | 5(10) |
| *B* | | | | | |
| 3.6(0.5) | | | | | |
| 2.3(0.4) | | | | | |
| 2.9(0.4) | | | | | |
| 2.7(0.4) | | | | | |

| $10^4\beta_{11}$ | $10^3\beta_{22}$ | $10^4\beta_{33}$ | $10^4\beta_{12}$ | $10^4\beta_{13}$ | $10^4\beta_{23}$ |
|---|---|---|---|---|---|
| 49(1) | 82(11) | 25(1) | −16(2) | −4(1) | −12(2) |
| 40(1) | 82(11) | 33(1) | 34(2) | 6(1) | 13(2) |
| 58(1) | 56(9) | 22(1) | −7(2) | 7(1) | −22(1) |
| 59(1) | 67(10) | 25(1) | 38(2) | 11(1) | −18(2) |
| 40(2) | 103(35) | 27(1) | 47(7) | −2(1) | −74(5) |
| 43(3) | 50(33) | 23(1) | 7(8) | 4(1) | −11(6) |
| 45(2) | 44(32) | 23(1) | 12(8) | 2(1) | −7(6) |
| 50(3) | 36(30) | 19(1) | −1(8) | 8(2) | 7(5) |
| 46(3) | 37(30) | 23(1) | 19(8) | 8(2) | 1(6) |
| 42(3) | 43(31) | 17(1) | −5(8) | −1(1) | 14(5) |
| 36(2) | 43(32) | 28(2) | 2(8) | 9(2) | 21(6) |

[b] Isotropic temperature factors (B) are in $A^2$; standard deviations are given in parentheses.

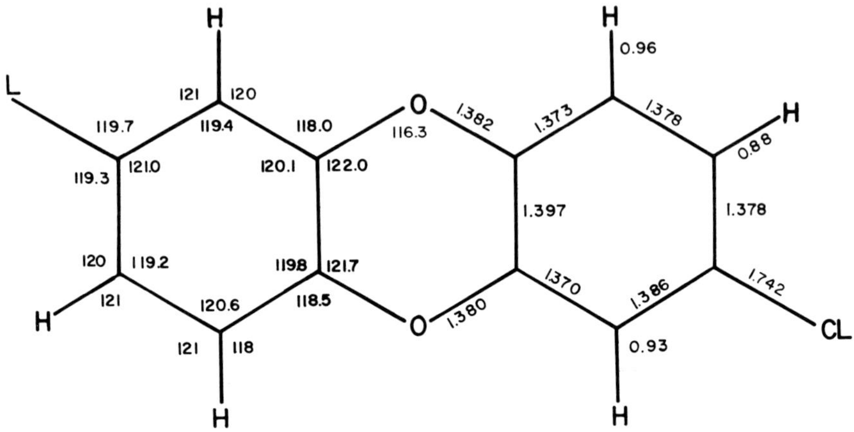


DCBD

| BONDS | | | ANGLES | |
|---|---|---|---|---|
| | C-C | ±0.005 Å | C-C-C | ±0.3° |
| | C-O | ±0.004 | C-C-O | ±0.3° |
| | C-Cl | ±0.004 | C-C-Cl | ±0.3° |
| | C-H | ±0.04 | C-C-H | ±2° |
| | | | C-O-C | ±0.3° |

*Figure 1. Bond distances and angles for 2,7-dichlorodibenzo-*p*-dioxin. The molecule is situated on an inversion center.*

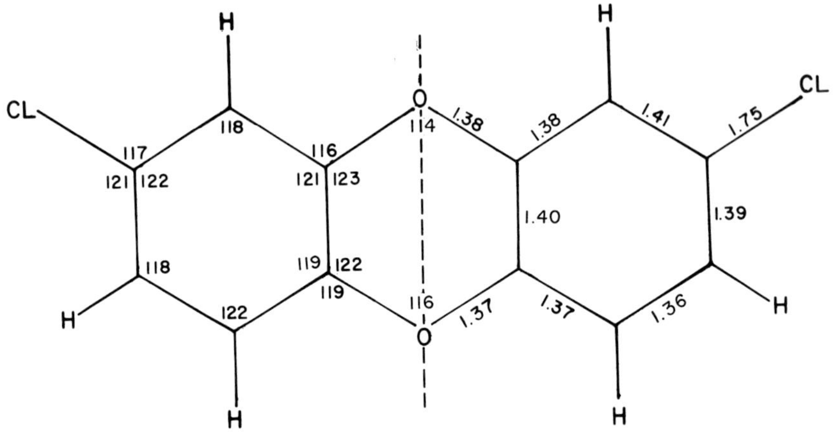


2,8-DCBD

| BONDS | | | ANGLES | |
|---|---|---|---|---|
| | C-C | ±0.01-0.02 Å | C-C-C | ±1.3° |
| | C-O | ±0.01 | C-C-O | ±1.3° |
| | C-CL | ±0.01 | C-C-CL | ±1.3° |
| | | | C-O-C | ±1.5° |

Site Symmetry *m*

*Figure 2. Bond distances and angles for 2,8-dichlorodibenzo-*p*-dioxin. The molecule is located on a crystallographic mirror plane.*

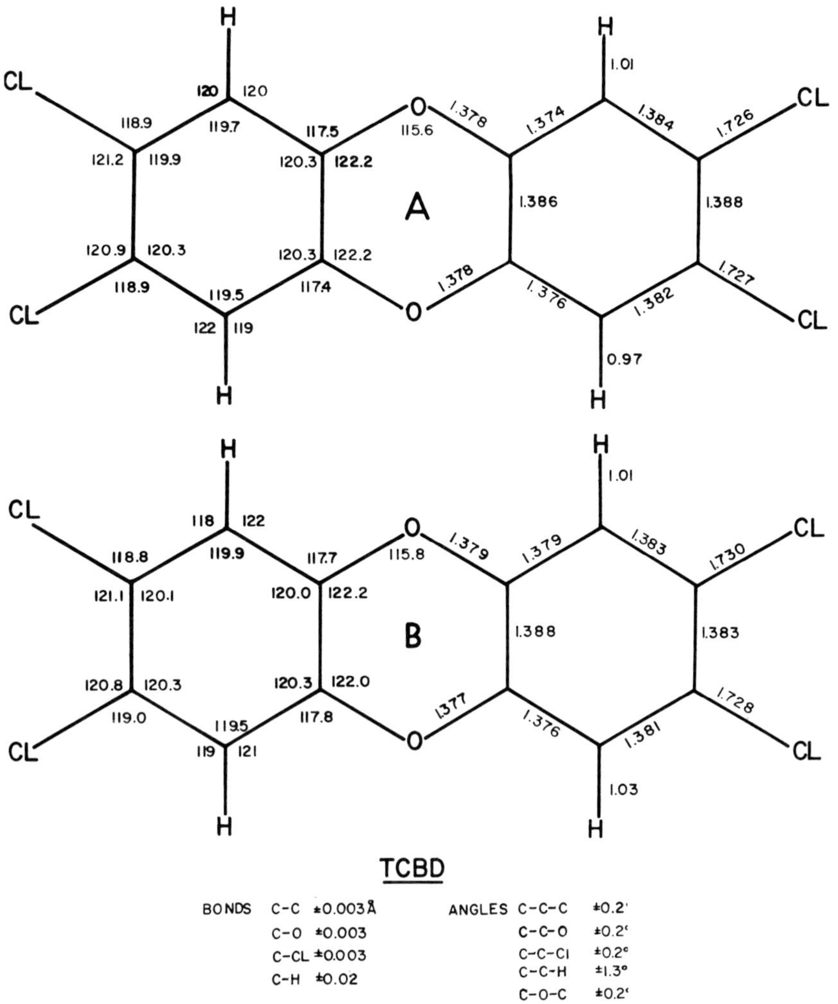


*Figure 3. Bond distances and angles for the two independent molecules in 2,3,7,8-tetrachlorodibenzo-*p*-dioxin. Both molecules are located on crystallographic inversion centers.*

X-ray powder diffraction, used wherever possible with x-ray single-crystal data, is a very convenient method for identifying samples of chlorinated dioxins and in particular for giving information on the isomers present. Some of the traditional methods for identifying organic molecules, such as infrared and mass spectrometry, are of limited utility in identifying isomers of the chlorodioxins. Nuclear magnetic resonance methods have been hindered by solubility problems; they also are less suitable for working with the very minute samples that can be analyzed

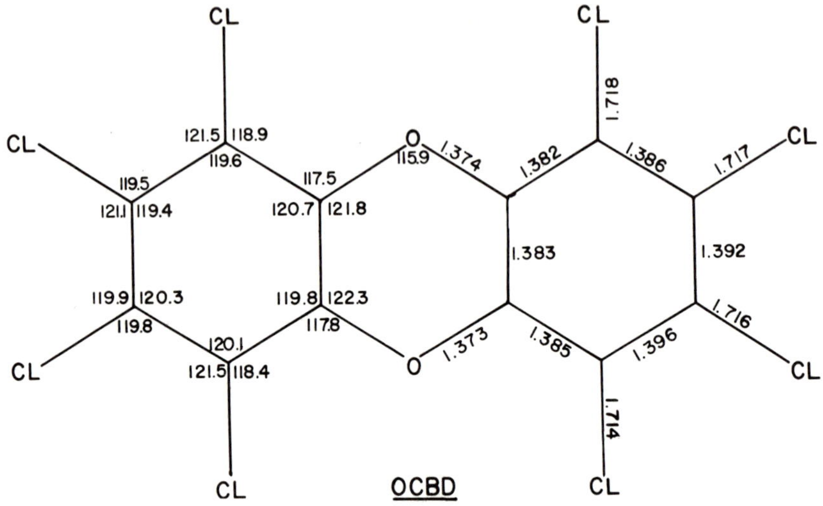


| BONDS | C-C | ±0.005 Å | ANGLES | C-C-C | ±0.4° |
|---|---|---|---|---|---|
| | C-O | ±0.004 | | C-C-O | ±0.3° |
| | C-CL | ±0.004 | | C-C-CL | ±0.3° |
| | | | | C-O-C | ±0.3° |

*Figure 4. Bond distances and angles for octachlorodibenzo-*p*-dioxin. The molecule is located on an inversion center.*

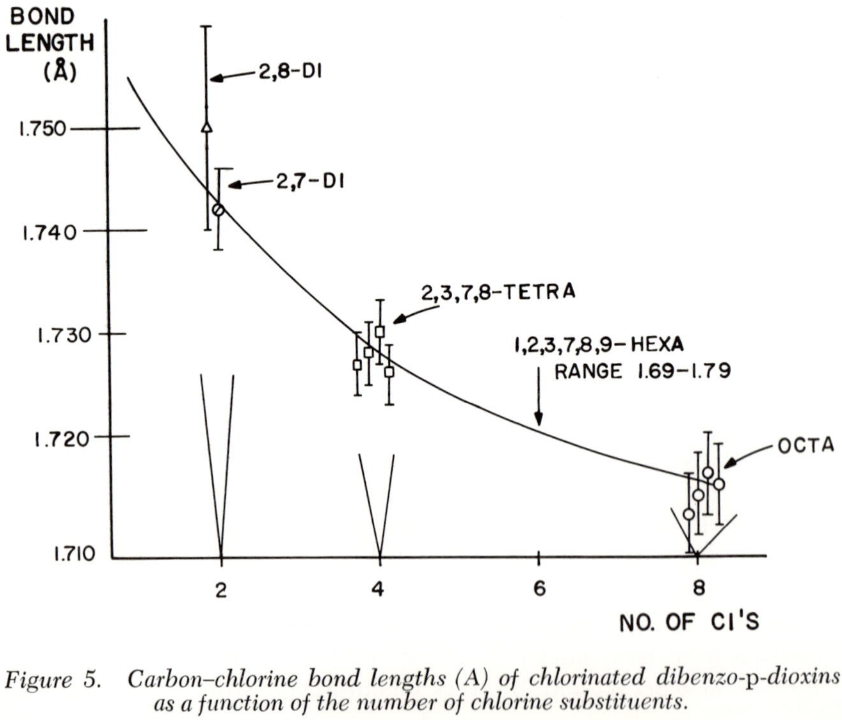


*Figure 5. Carbon–chlorine bond lengths (A) of chlorinated dibenzo-*p*-dioxins as a function of the number of chlorine substituents.*

**Table III. X-ray Powder Diffraction Data for 2,3,7,8-Tetrachlorodibenzo-*p*-dioxin[a] and 2,7-Dichlorodibenzo-*p*-dioxin[b]**

*TCBD[c]*

| *d(hkl)* | *I/I(o)* | *hkl* |
|---|---|---|
| 8.728 | 3 | 011 |
| 8.130 | 4 | 011 |
| 6.369 | 23 | 012 |
| 5.901 | 24 | 012 |
| 4.984 | 2 | 020 |
| 4.851 | 3 | 021 |
| 4.739 | 6 | 013 |
| 4.639 | 3 | 021 |
| 4.448 | 6 | 013 |
| 4.343 | 15 | 022 |
| 4.046 | 10 | 022 |
| 3.759 | 10 | 100 |
| 3.736 | 7 | 101, 023 |
| 3.578 | 50 | 101, 110 |
| 3.537 | 2 | 111, 014 |
| 3.456 | 60 | 110, 111 |
| 3.330 | 100 | 112 |
| 3.296 | 38 | 111, 031 |
| 3.266 | 23 | 102 |
| 3.188 | 25 | 103, 031 |
| 3.180 | 22 | 024 |
| 3.035 | 50 | 113, 015, 112 |
| 3.003 | 3 | 121 |
| 2.953 | 17 | 024 |
| 2.916 | 13 | 103 |
| 2.892 | 5 | 015, 122, 033 |
| 2.866 | 4 | 113 |
| 2.850 | 4 | 122 |

*2,7-DCBD*

| *d(hkl)* | *I/I(o)* | *hkl* |
|---|---|---|
| 9.9 | 13 | 001 |
| 6.55 | 25 | 010 |
| 5.00 | 63 | 011 |
| 4.39 | 5 | 012 |
| 3.56 | 63 | 110 |
| 3.46 | 25 | 111 |
| 3.26 | 100 | 101, 020, 111 |
| 3.01 | 10 | 022 |
| 2.92 | 20 | 021, 112 |
| 2.76 | 5 | 112, 120 |
| 2.57 | 8 | 113 |
| 2.48 | 8 | 014, 004 |
| 2.38 | 5 | 122 |

**Table III. Continued**

*2,7-DCBD*

| d(hkl) | I/I(o) | hkl |
|---|---|---|
| 2.30 | 5 | 104, 112 |
| 2.21 | 5 | 031, 114 |
| 2.12 | 18 | 114, 123 |
| 2.07 | 5 | 130, 121 |
| 2.03 | 5 | 031 |

[a] Obtained on a 115-mm diameter AEG (Allgemeine Elektrizitats–Gesellschaft) Guinier camera using Seeman–Bohlin focusing and Cu K$\alpha$, radiation ($\lambda$ 1.5405 A).

[b] Data taken with a 143.2-mm diameter Debye–Scherrer camera using Cu K$\alpha$ radiation ($\lambda$ 1.5418 A).

[c] Forty additional lines to 1.603 A were recorded

by x-ray diffraction. The observed powder patterns for the 2,7-dichloro- and the 2,3,7,8-tetrachlorodibenzo-*p*-dioxins are given in Table III.

Several cases have been encountered that show the utility of x-ray powder diffraction as an analytical method for identifying dioxins. An attempt was made to synthesize 2,8-dichlorodibenzo-*p*-dioxin by heating the potassium salt of a trichlorinated 2-hydroxydiphenyl ether (*3*). Surprisingly, the major crystalline product under the initial reaction condi-

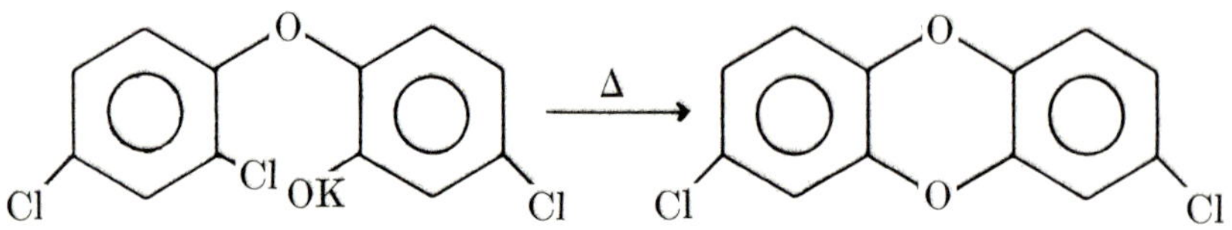


tions was 2,7-dichlorodibenzo-*p*-dioxin suggesting that cleavage of the ether and its subsequent reformation had occurred unexpectedly. Some 2,8-dioxin could also be detected in the reaction product. A second sample of 2,8-dioxin contained an unknown crystalline impurity which could be physically separated on the basis of crystal morphology.

In addition, an interesting, although negative, result has come from powder diffraction studies of the hexachloro compounds. We have examined Debye–Scherrer photographs of several samples known to contain predominantly hexachlorodibenzo-*p*-dioxins and have identified the patterns of at least three crystalline phases therein. (There are 10 possible isomers of hexachlorodibenzo-*p*-dioxin.) These patterns have been checked carefully against the calculated *d*-spacings and intensities of the 1,2,3,7,8,9-hexa isomer described by Cantrell, Webb, and Mabis (*1*) and also against an observed pattern supplied by Cantrell and believed to be from the low temperature phase of the same material. Yet to date we have not been able to detect the signature of either of these 1,2,3,7,8,9-hexachlorodioxin phases in any sample examined.

### *Acknowledgment*

The authors are pleased to acknowledge the contributions of W. W. Muelder and O. Aniline, who provided the samples and much stimulating discussion.

### *Literature Cited*

1. Cantrell, J. S., Webb, N. C., Mabis, A. J., *Acta Crystallogr.* (1969) **B25,** 150.
2. Schomaker, V., Trueblood, K. N., *Acta Crystallogr.* (1968) **B24,** 63.
3. Aniline, O., private communication (1971).

RECEIVED February 8, 1972.

# 4

# Thermal Chemistry of Chlorinated Phenols

H. G. LANGER, T. P. BRADY, L. A. DALTON, T. W. SHANNON, and P. R. BRIGGS

The Dow Chemical Co., Eastern Research Laboratory, Wayland, Mass. 01778

*Thermal properties of several chlorinated phenols and derivatives were studied by differential thermal analysis and mass spectrometry and in bulk reactions. Conditions which might facilitate the formation of stable dioxins were emphasized. No two chlorinated phenols behaved alike. For a given compound the decomposition temperature and rate as well as the product distribution varied considerably with reaction conditions. The phenols themselves seem to pyrolyze under equilibrium conditions slowly above 250°C. For their alkali salts the onset of decomposition is sharp and around 350°C. The reaction itself is exothermic. Preliminary results indicate that heavy ions such as cupric ion may decrease the decomposition temperature.*

The literature on formation of halodibenzo-*p*-dioxins is vague and sometimes confusing. Most of the information available is summarized in an article by M. Kulka (*1*) and deals with the formation of octachlorodibenzo dioxin from pentachlorophenol. We have investigated the thermal properties of several chlorinated phenols and derivatives by differential thermal analysis and mass spectrometry and in bulk reactions. The reaction conditions were chosen to favor the dioxin formation.

Of the chlorinated phenols pentachlorophenol is the most likely to produce a dioxin. Its differential thermogram, however, (Figure 1) reveals no decomposition. A solid state transition at 75°C is followed by melting below 200°C and vaporization above 300°C. Only after prolonged heating in bulk and at temperatures above 200°C could octachlorodibenzodioxin be identified in the tarry residue. If the compound is heated in a sealed capillary, conversions of ~50% are obtained clearly indicating an equilibrium (Figure 2). Here the area of the melting endotherm is used to measure the progress of the reaction. Only a

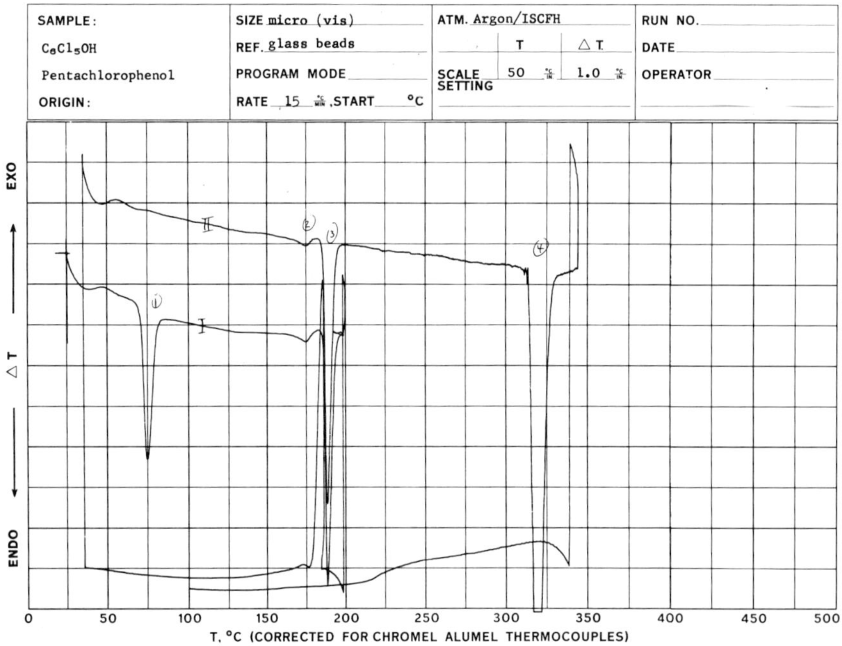


*Figure 1. Differential thermogram of pentachlorophenol*

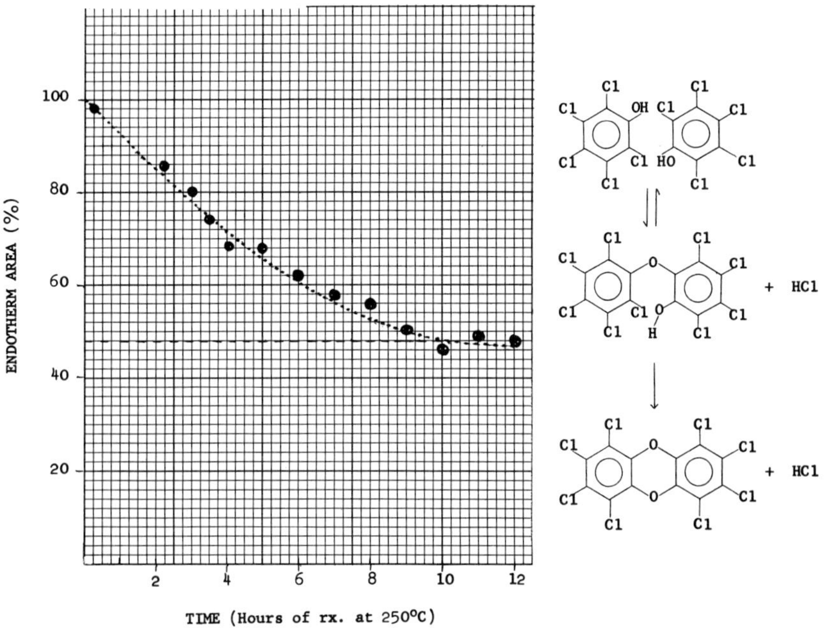


*Figure 2. Thermal decomposition of $C_6Cl_5OH$*

small amount of dioxin was detected in the sample at the end of the experiment—the main product being the chlorinated phenoxyphenol, indicating an equilibrium for the first step of the condensation reaction.

A drastically different reaction is indicated on the thermogram of sodium pentachlorophenate (Figure 3) as a strongly exothermic reaction at ~360°C. Upon cooling, essentially pure octachloro-*p*-dibenzodioxin crystallized.

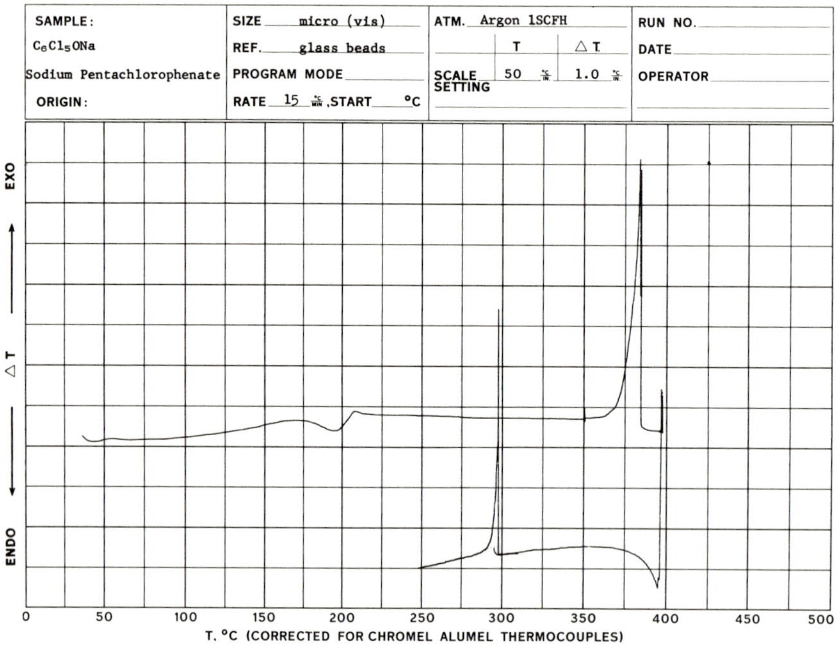


*Figure 3. Differential thermogram of sodium pentachlorophenate*

For 2,3,4,6-tetrachlorophenol hardly any discoloration was detected above 200°C; instead the sample vaporized rapidly. Its sodium salt reacted similarly to sodium pentachlorophenate except that its exothermic decomposition (Figure 4) is less clearly defined; no crystallization occurred on cooling, and the yield of hexachlorodioxins was much lower, the remaining products being higher molecular weight materials. The hexachlorodioxins were identified by gas–liquid chromatography as two isomers in a ratio of 35:65.

For lower chlorinated phenates the reactions become even more complicated. The phenols vaporize without decomposition, and the sodium salts react in the melt (Figure 5), as shown for sodium 2,4-dichlorophenate. The relative amount of dioxins produced is lower in favor of more highly condensed material.

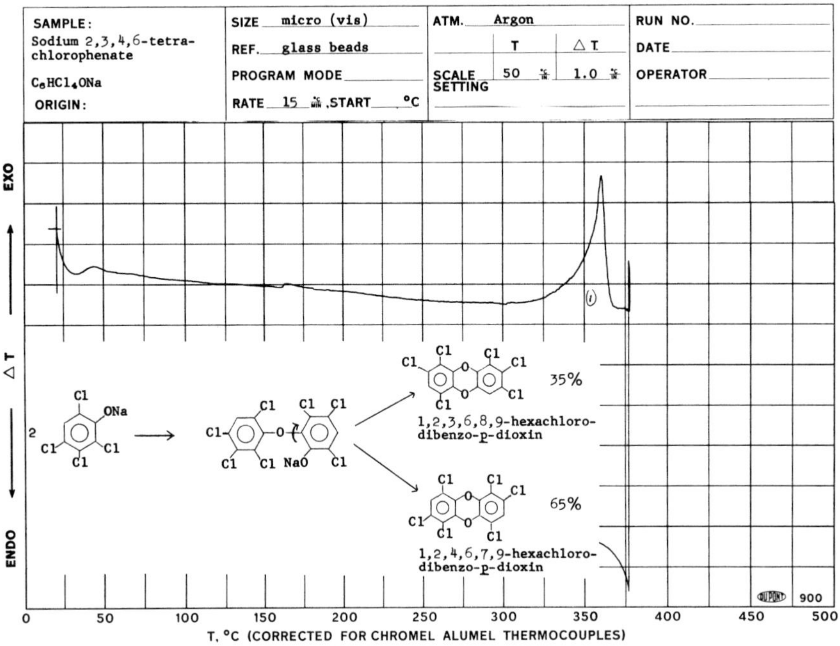


*Figure 4. Differential thermogram of sodium 2,3,4,6-tetrachlorophenate*

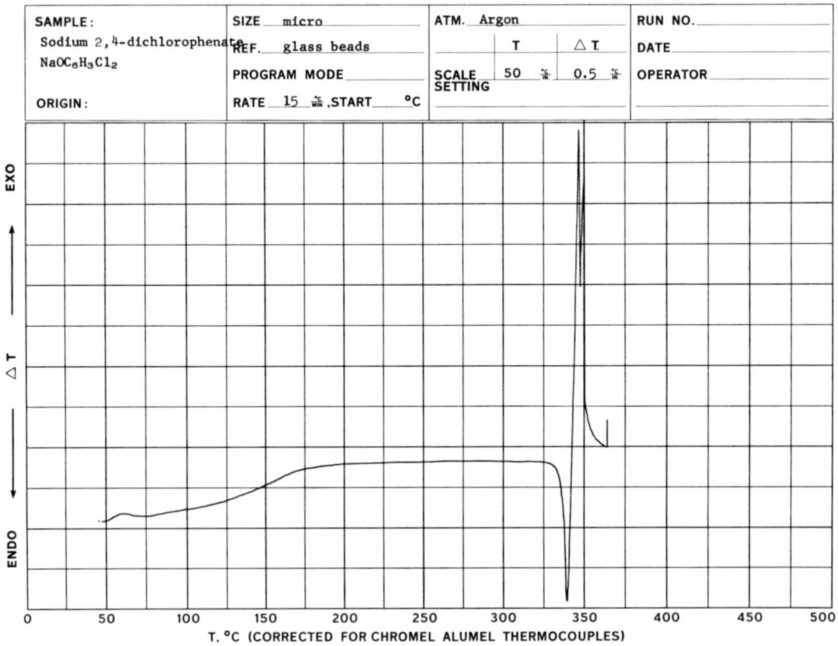


*Figure 5. Differential thermogram of sodium 2,4-dichlorophenate*

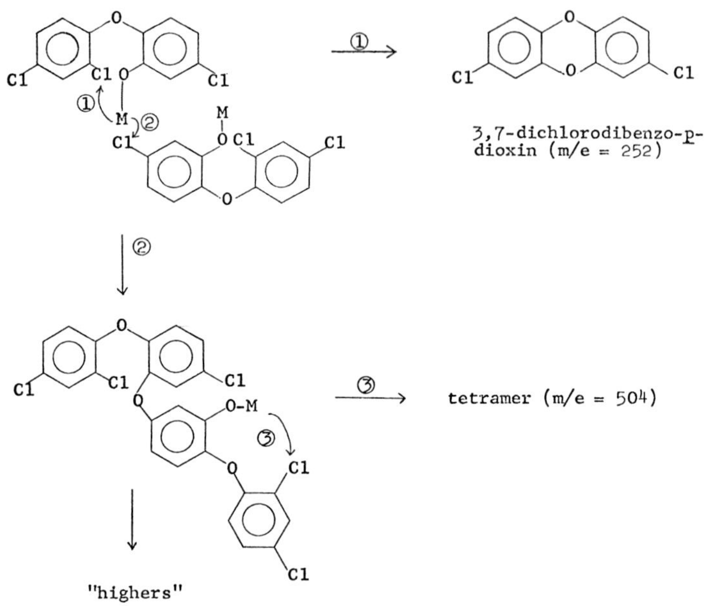


*Figure 6. Irgasan, 2-(2,4-dichlorophenoxy)-5-chlorophenol*

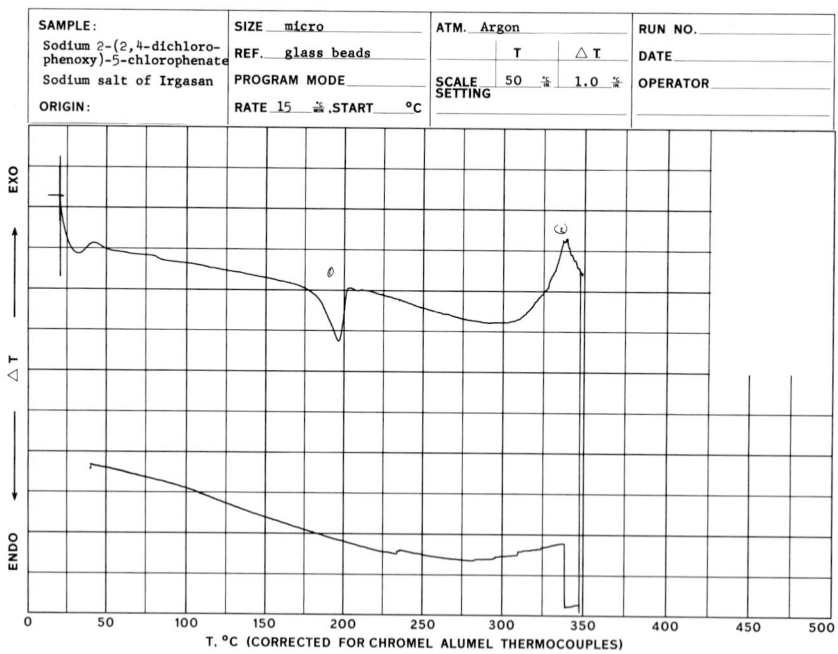


*Figure 7. Differential thermogram of sodium 2-(2,4-dichlorophenoxy)-5-chlorophenate*

**Table I. Mass Spectrometric Analysis of Decomposition Residues**

| m/e | Relative Intensity, % | Species |
|---|---|---|
| *Sodium 2,4-dichlorophenate ($C_6H_3Cl_2ONa$)* | | |
| 252 | 100 | dioxin (2Cl) |
| 288 | 12 | phenoxyphenol (3Cl) |
| 378 | 10 | trimer (3Cl) |
| 414 | 5 | phenoxyphenoxyphenol (4Cl) |
| 504 | 0.5 | tetramer (4Cl) |
| 540 | 0.3 | phenoxy(phenoxy)$_2$ phenol (5Cl) |
| *Sodium "Irgasan" ($C_6H_3Cl_2OC_6H_3ClONa$)* | | |
| 252 | 100 | dioxin (2Cl) |
| 378 | 10 | trimer (3Cl) |
| 414 | 20 | phenoxyphenoxyphenol (4Cl) |
| 504 | 16 | tetramer (4Cl) |
| 540 | 12 | phenoxy(phenoxy)$_2$ phenol (5Cl) |

The compound 2-(2,4-dichlorophenoxy)-5-chlorophenol can be considered an intermediate for the 3,7-dichlorodibenzo-*p*-dioxin formation (Figure 6). One might expect that condensation of its sodium salt by Reaction I leads primarily to the dioxin rather than to higher condensed materials *via* Pathway II. Differential thermal analysis (Figure 7) shows an ill-defined exothermic reaction starting considerably above the melting temperature. The product, however, is not the expected pure dioxin but a complicated reaction mixture (Table I). The amount of higher condensed materials is surprisingly high compared with that from sodium-2,4-dichlorophenate.

**Table II. Dioxin Formation *vs.* "Highers"**

| m/e | Relative Intensity, % | Species |
|---|---|---|
| *Sodium Pentachlorophenate* | | |
| 456 | 100 | dioxin (8Cl) |
| 684 | 2 | trimer (12Cl) |
| 912 | 0.3 | tetramer (16Cl) |
| *Sodium 2,3,4,6-tetrachlorophenate* | | |
| 388 | 100 | dioxin (6Cl) |
| 582 | 3 | trimer (9Cl) |
| *Sodium 2,4-dichlorophenate* | | |
| 252 | 100 | dioxin (2Cl) |
| 378 | 10 | trimer (3Cl) |
| 504 | 1 | tetramer (4Cl) |
| *Sodium "Irgasan"* | | |
| 252 | 100 | dioxin (2Cl) |
| 378 | 10 | trimer (3Cl) |
| 504 | 16 | tetramer (4Cl) |

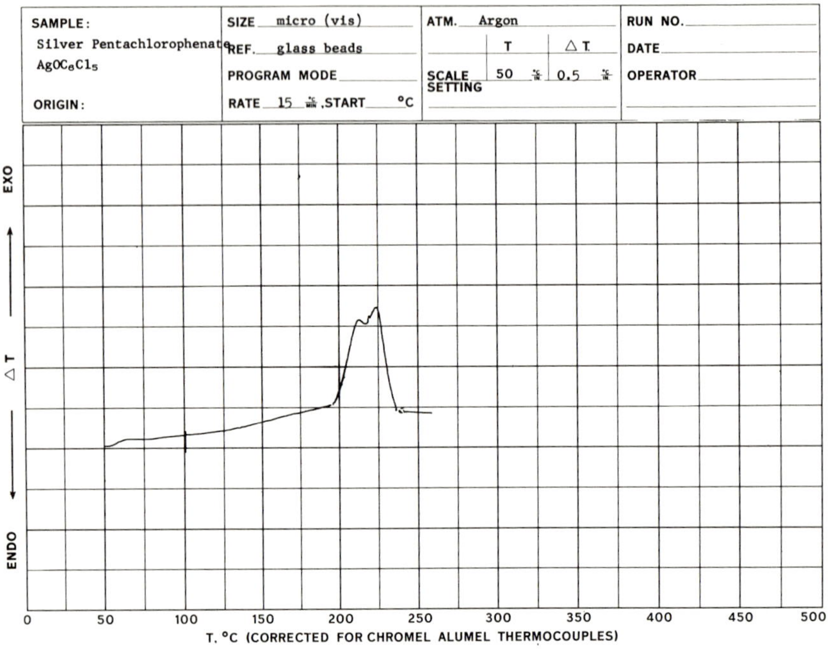


*Figure 8. Differential thermogram of silver pentachlorophenate*

In summary thermal decomposition of chlorinated phenols does not generally lead to dioxins. There are, however, several conditions which by themselves or combined would favor dioxin formation. First, of all chlorinated phenols either in bulk or in solution, only pentachlorophenol produced measurable amounts of dioxin. Secondly (Table II), only sodium salts in solid state reactions produced dioxins in reasonable yields. In contrast, the silver salt of pentachlorophenol (Figure 8) undergoes an exothermic decomposition at considerably lower temperatures and produced only higher condensed materials. No dioxin was detected.

Thus, it seems that dioxins are only formed if the intermediate phenoxyphenol can be forced and held in a special molecular configuration to avoid the various competitive reactions leading to different products.

## *Literature Cited*

1. Kulka, M., *Can. J. Chem.* (1961) **39**, 1973.

Received February 8, 1972.

# 5

# Cation Radicals of Chlorinated Dibenzo-*p*-dioxins

G. C. YANG and A. E. POHLAND

Division of Chemistry and Physics, Food and Drug Administration,
U. S. Department of Health, Education, and Welfare, Washington, D. C. 20204

*The electron spin resonance and visible absorption spectra of a series of chlorinated dibenzo-*p*-dioxin cation radicals containing one to eight chlorine atoms are obtained by dissolution in trifluoromethanesulfonic acid. An oxidizing agent or UV irradiation is needed only to generate the cation radicals of the highly chlorinated isomers. A well degassed sample of the dibenzo-*p*-dioxin cation radical gives narrow linewidths (<40 moe), extremely long lifetime (>1 year) with* g = *2.0038 and well resolved* $C^{13}$ *hyperfine lines. The optical absorption band occurs at 655 nm. Detailed data on the visible and ESR spectra of the cation radicals from a series of chlorinated dibenzo-*p*-dioxins are also tabulated and discussed.*

Compounds having the dibenzo-*p*-dioxin nucleus give a blue to blue-green color when dissolved in concentrated sulfuric acid with oxidizing agents, such as $H_2O_2$ and $KNO_3$ (*1*).

This color transformation has been observed in dibenzo-*p*-dioxin (Structure I) and in its bromo, chloro, nitro, methyl, and ethyl derivatives; in addition, the observed electron spin resonance (ESR) signals indicated the presence of paramagnetic species (*2, 3*). This phenomenon has been attributed to the formation of cation radicals in acid solution.

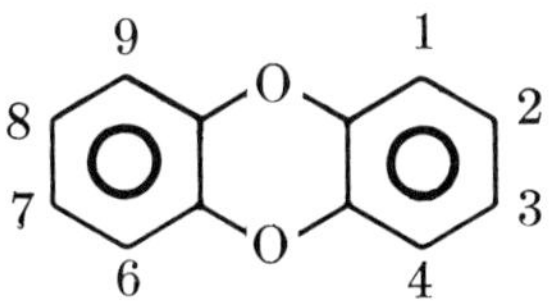


(I)

Studies were made on a series of chlorinated dibenzo-*p*-dioxin cation radicals in trifluoromethanesulfonic acid (TFMS acid). TFMS acid was selected instead of sulfuric acid for the following reasons: (1) cation radical lifetimes were found to increase many fold, (2) microwave dielectric loss was less severe, and (3) oxidizing agents were not needed for the less highly chlorinated dibenzo-*p*-dioxins, while for the more highly chlorinated derivatives, the cation radicals were generated through UV irradiation or by addition of an oxidizing agent. Shine *et al.* (*4*) proposed a mechanism for the formation of cation radicals of thianthrene in concentrated sulfuric acid. A similar mechanism is proposed here for the formation of cation radicals of dibenzo-*p*-dioxin in concentrated sulfuric acid.

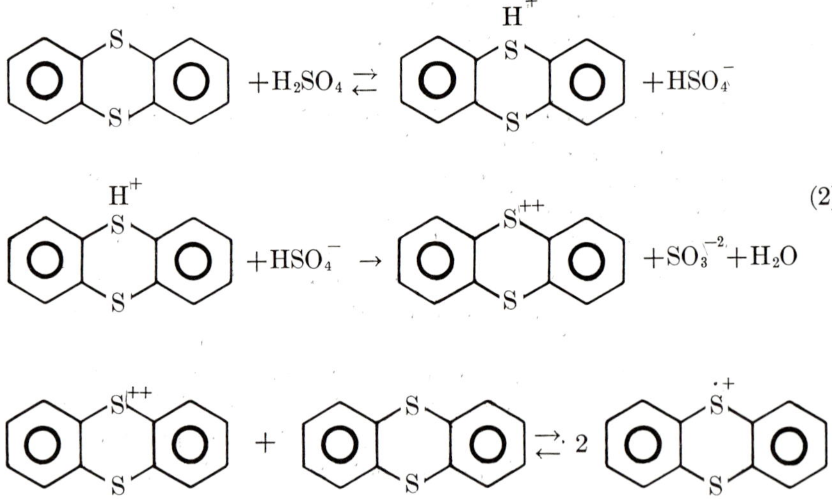


(2)

In TFMS acid, radical cation formation was almost negligible when the TFMS acid was degassed before mixing with the dioxins, and only after some time was any appreciable radical formation observed. The following mechanism, with dissolved oxygen and/or UV irradiation serving as oxidizing agent, is proposed.

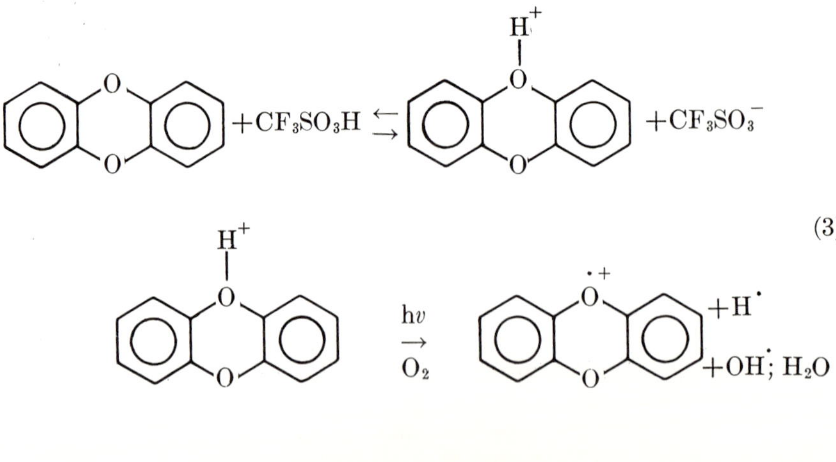


(3)

**Table I. Absorption Maxima of Chlorinated Dibenzo-*p*-dioxins in TFMS Acid**

| | *Absorption Peak, nm* |
|---|---|
| dibenzo-*p*-dioxin | 655.0 |
| 1-chloro | 708.0 |
| 2-chloro | 674.5 |
| 2,7-dichloro | 720.0 |
| 2,8-dichloro | 762.0 |
| 2,3-dichloro | 742.0 |
| 1,2,4-trichloro | 706.0 |
| 1,3,6,8-tetrachloro | 733.0 |
| 2,3,7,8-tetrachloro | 845.0 |
| 1,2,3,4-tetrachloro | 758.0 |
| 1,2,3,4,7-pentachloro | 790.0 |
| 1,2,4,6,7,9-hexachloro | 750.0 |

The visible absorption maxima for a series of chlorinated dibenzo-*p*-dioxins in TFMS acid are tabulated in Table I. The visible absorption spectra of these compounds vary with the position and number of chlorine atoms. In general, a bathochromic shift was observed as more chlorine atoms were attached to the dibenzo-*p*-dioxin nucleus.

Adding an oxidizing agent to the acid solvent caused the dioxin cation radical resonance line to broaden, usually with the loss of hyperfine splittings. However, solubility, especially for the highly chlorinated dioxins, improves greatly with an oxidizing agent present. The *g*-factors and linewidths for a series of chlorinated dibenzo-*p*-dioxins are tabulated in Table II with Structure I as a standard.

**Table II. *g*-Factors and Linewidth for the Chlorinated Dibenzo-*p*-dioxins in TFMS Acid with $KNO_3$[a]**

| | *g-Factor* | $\Delta H$ *(oe)* | *Linewidth (oe)* |
|---|---|---|---|
| dibenzo-*p*-dioxin | 2.0038 | — | .090 |
| 1-chloro | 2.0027 | 1.82 | 3.26 |
| 2-chloro | 2.0034 | .67 | 2.13; 4.91 |
| 2,3-dichloro | 2.0018 | 3.31 | 4.34 |
| 2,7-dichloro | 2.0026 | 2.17 | 3.33 |
| 2,8-dichloro | 2.0024 | 2.31 | 2.45 |
| 1,2,4-trichloro | 2.0029 | 1.50 | 4.34 |
| 1,2,3,4-tetrachloro | 2.0017 | 3.47 | 4.23 |
| 2,3,7,8-tetrachloro | 2.0020 | 3.10 | 3.16 |
| 1,3,6,8-tetrachloro | 2.0025 | 2.13 | 3.79 |
| 1,2,3,4,7-pentachloro | 2.0019 | 3.18 | 3.42 |
| 1,2,3,4,7,8-hexachloro | 2.0017 | 3.60 | 3.68 |
| 1,2,4,6,7,9-hexachloro | 2.0024 | 2.29 | 4.86 |
| 1,2,3,4,6,7,8,9-octachloro | 2.0016 | 3.76 | 3.39 |

[a] $\Delta H = H_1 - H_0$ where $H_0$ is dibenzo-*p*-dioxin and $H_1$ is chlorinated dibenzo-*p*-dioxin; $_k \nu_0 = 9.5075$ $KMH_z$.

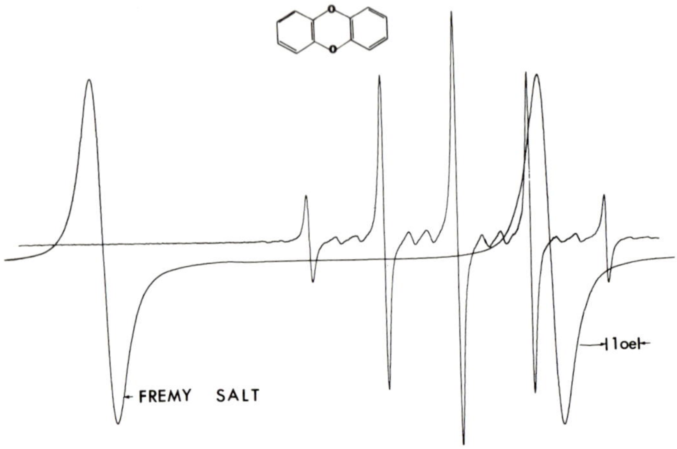


*Figure 1. ESR spectrum of dibenzo-*p*-dioxin in TFMS acid*

When the chlorinated dibenzo-*p*-dioxins were dissolved in TFMS acid without any oxidizing agent and after the sample tube was carefully degassed, much narrower linewidths were observed, usually with resolved hyperfine splittings and greatly prolonged radical lifetimes. Figure 1 shows the spectrum of the unsubstituted dibenzo-*p*-dioxin in TFMS acid with Fremy salt as a calibration standard. Figure 2 shows the same spectrum on an expanded scale. The five-line spectrum with intensity

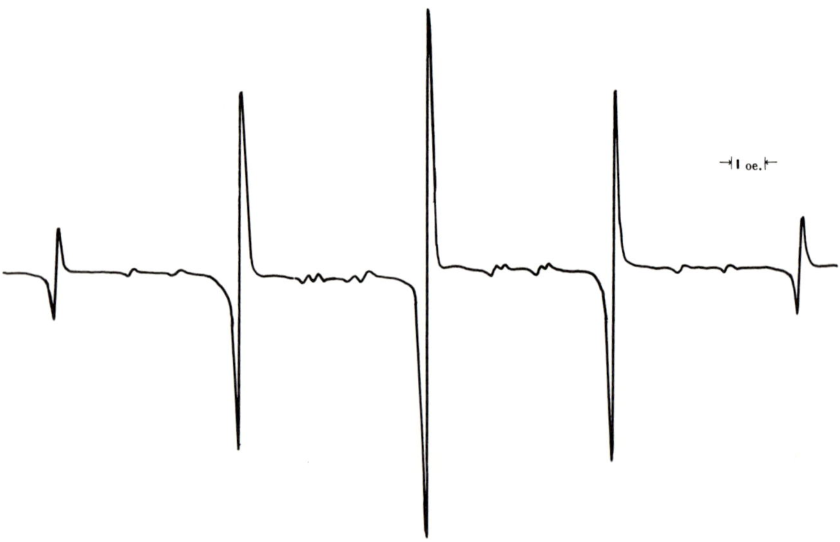


*Figure 2. ESR spectrum of dibenzo-*p*-dioxin in TFMS acid with dilute concentration*

ratio of 1:4:6:4:1 may be attributed to coupling of the unpaired electron with a set of four equivalent protons. The protons responsible for the observed hyperfine splitting have been assigned to the 2, 3, 7, and 8 positions while the 1, 4, 6, and 9 positions offer negligible spin densities similar to the assignment made in the cation radicals of thianthrene (*4*). This conclusion is in agreement with the chemical reactivity toward electrophilic reagents at these sites (*5*) and with the INDO–MO calculation of spin densities at these sites (*6*). In concentrated sulfuric acid, Tozer and Tuck (*7*) reported a slightly different value for the *g*-factor and hyperfine constant (*see* Table III).

**Table III. Comparison of Measured ESR Parameters of Dibenzo-*p*-dioxin in TFMS and Sulfuric Acids**

| *Parameters* | *TFMS* | $H_2SO_4$ [a] | *Hyperfine Coupling Constant* | |
|---|---|---|---|---|
| *g*-Factor | 2.0038 | 2.0034 | — | — |
| A (oe) | 2.178 | 2.52, 2.13 [b] | 2,3,7,8 | protons |
| A′ (oe) | <.3 | <.5 | 1,4,6,9 | protons |
| $A_{C^{13}}$ (oe) | 1.650 | 1.77 | 2,3,7,8 | carbons |
| $A'_{C^{13}}$ (oe) | 2.708 | 3.27 | 1,4,6,9 | carbons |
| Linewidth (oe) | .09 | .17 | — | — |

[a] Data of Tozer and Tuck, (*7*).
[b] Unpublished data of Yang and Pohland.

The 1-chloro- and 2-chlorodibenzo-*p*-dioxins, which readily dissolved in TFMS acid, formed cation radicals without UV irradiation or the addition of oxidizing agents. With the exception of broader resonance lines, Figure 3 shows that the five-line pattern observed with 1-chlorodibenzo-p-dioxin is similar to that of the unsubstituted dibenzo-*p*-dioxin. Apparently, protons at the 2, 3, 7, and 8 positions became less equivalent

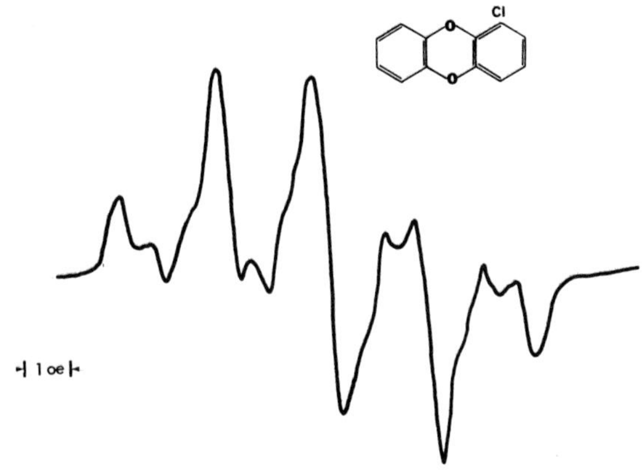


*Figure 3. ESR spectrum of 1-chlorodibenzo-*p*-dioxin in TFMS acid*

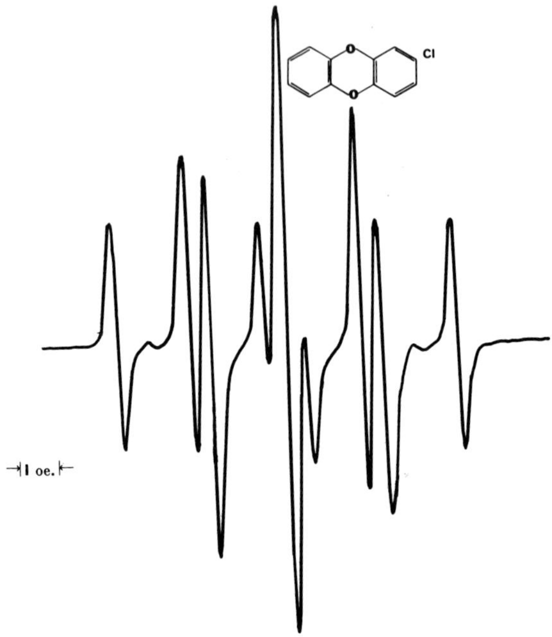


*Figure 4. ESR spectrum of 2-chlorodibenzo-*p*-dioxin in TFMS acid*

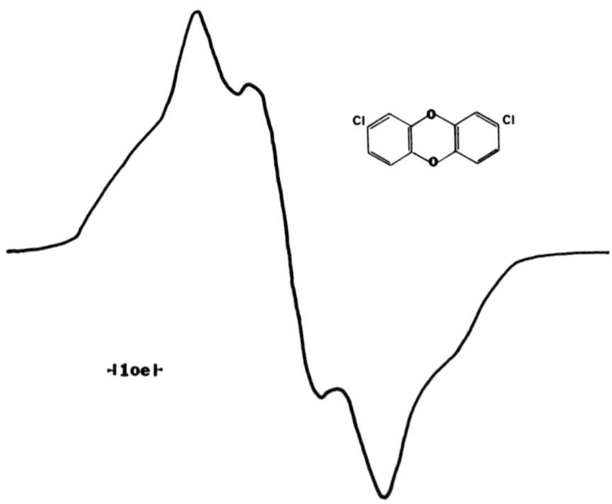


*Figure 5. ESR spectrum of 2,8-dichlorodibenzo-*p*-dioxin in TFMS acid*

because of the substitution of a chlorine atom in the 1 position, resulting in further splitting and concomitant broadening of the resonance lines. For 2-chlorodibenzo-*p*-dioxin (Figure 4), a resolved nine-line spectrum was observed. This spectrum can be interpreted in terms of hyperfine splittings arising from two sets of two equivalent protons. Since symmetry considerations do not indicate this configuration, ESR studies of other 2-substituted derivatives as well as MO calculations are being made to clarify this situation.

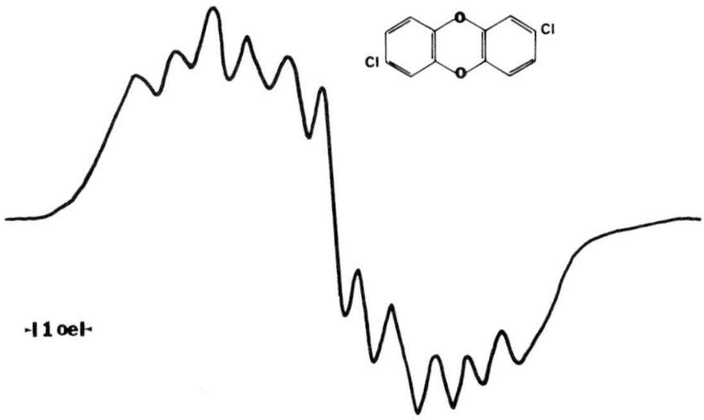


*Figure 6. ESR spectrum of 2,7-dichlorodibenzo-*p*-dioxin in TFMS acid*

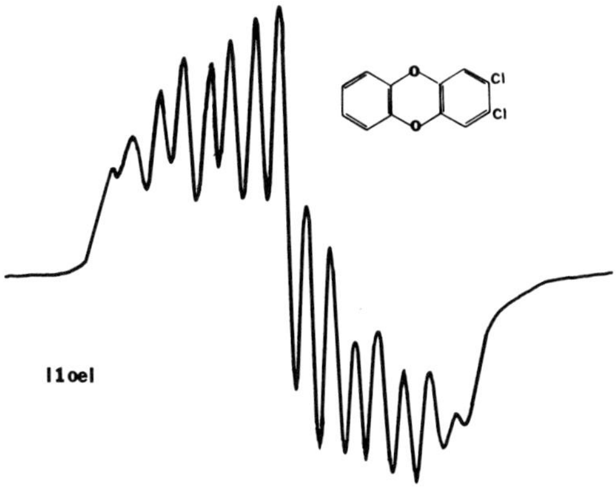


*Figure 7. ESR spectrum of 2,3-dichlorodibenzo-*p*-dioxin in TFMS acid*

Four dichloro isomers, the 1,6-, 2,3-, 2,7-, and 2,8-dichlorodibenzo-*p*-dioxins, were studied. These compounds also dissolve in TFMS acid, forming cation radicals in the absence of oxidizing agents or UV irradiation. The 2,8-isomer (Figure 5) exhibited a three-line spectrum, in agreement with the two equivalent protons in the 3,6 positions. The 2,7-isomer should also exhibit a three-line spectrum, similar to the 2,8-

isomer; instead, 11 resolved hyperfine lines were detected (Figure 6). Similarly, the 2,3-isomer (Figure 7) exhibited a 15-line spectrum. Using a computer spectrum simulation program, the additional hyperfine splittings can be explained. Interactions between the unpaired electron spin and various chlorine nuclei and the additional line broadenings resulting from those chlorine isotopes possessing large quadrupole moments must be considered (7). The 1,6-isomer (Figure 8) exhibited a five-line spectrum, similar to that of Structure I, emphasizing again that the 1, 4, 6, and 9 positions indeed possess low spin density.

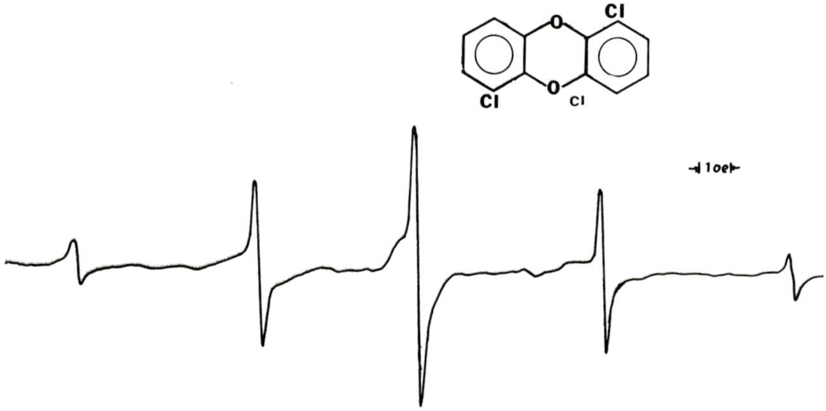


*Figure 8. ESR spectrum of 1,6-dichlorodibenzo-*p*-dioxin in TFMS acid*

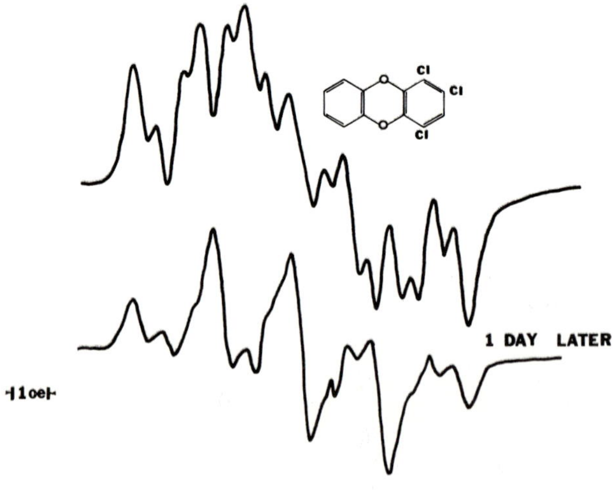


*Figure 9. ESR spectrum of 1,2,4-trichlorodibenzo-*p*-dioxin in TFMS acid after UV irradiation*

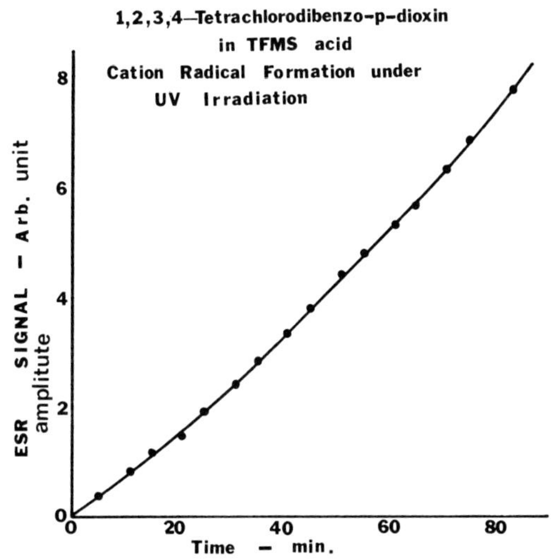


*Figure 10. ESR signal amplitude growth with time under constant irradiation*

1,2,4-Trichlorodibenzo-*p*-dioxin was sensitive to photodecomposition. When this compound was irradiated in TFMS acid by UV light, the upper spectrum in Figure 9 was obtained. However a new spectrum appeared after standing for a day, identical to the spectrum of the 1-chloro derivative. Further photochemical studies with this compound are being made.

*Figure 11. ESR spectrum of 1,2,3,4-tetrachlorodibenzo-*p*-dioxin in TFMS acid after UV irradiation*

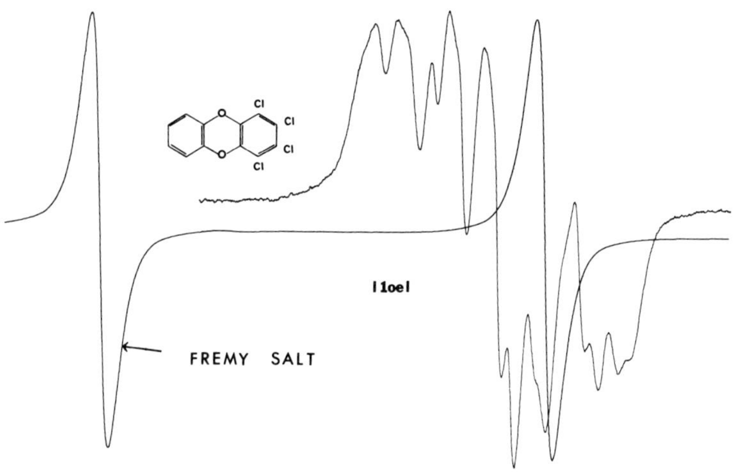

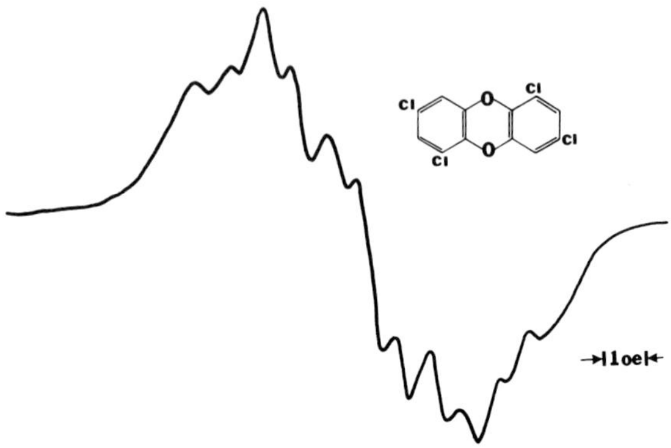


*Figure 12. ESR spectrum of 1,3,6,8-tetrachlorodibenzo-*p*-dioxin in TFMS acid after UV irradiation*

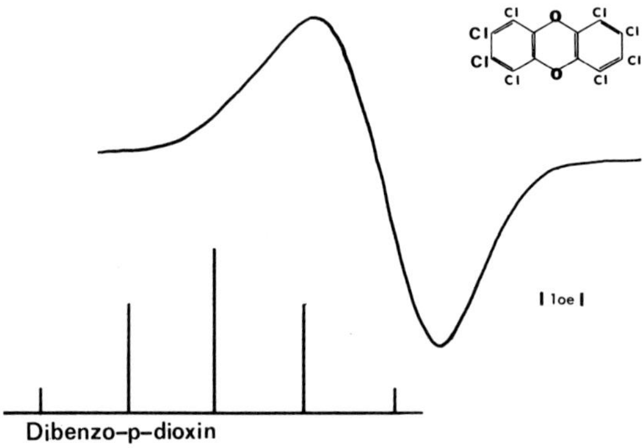


*Figure 13. ESR spectrum of octachlorodibenzo-*p*-dioxin in TFMS acid*

Three isomeric tetrachlorodibenzo-*p*-dioxins were studied. All were insoluble in TFMS acid. To dissolve these compounds and form cation radicals, UV irradiation was necessary. The 1,2,3,4-tetrachloro compound was particularly sensitive to UV irradiation, and as a solid, even turned pink when exposed to ordinary fluorescent light. When subjected to constant UV irradiation, radical ions were induced rapidly. The change in the cation radical concentration was monitored by the ESR signal as illustrated in Figure 10. To determine whether the tetrachloro isomer had been converted to lower chlorinated derivatives after UV irradiation, the dissolved dioxin was then poured into ice water and recovered. The GLC retention time of the recovered dioxin was unchanged; in addition, no new GLC peaks were observed. Moreover, the ESR spectrum (*see* Figure 11) for the recovered material was not altered between widely separated irradiation times (*ca.* 5 to 6 hours). There are 14 hyperfine lines for the 1,3,6,8-tetrachlorodibenzo-*p*-dioxin (Figure 12). On the other hand, the toxic 2,3,7,8-isomer exhibited only one broad line. For the fully chlorinated octachlorodibenzo-*p*-dioxin, no hyperfine structures were detected (*see* Figure 13).

### *Summary*

We have studied some of the properties of a series of chlorinated dibenzo-*p*-dioxin radical cations. Both the *g*-factors and the wavelengths of the visible absorption peaks of the cation radicals vary with the number and the position of the chlorine atoms, undoubtedly because of the electron withdrawing power of chlorine atoms. Unfortunately, line broadening, also caused by the chlorine atoms, made the exact interpretation of the hyperfine splitting extremely difficult. Both the photogeneration of the radical ions and their photodecomposition in TFMS acid are interesting and need further study. With two criteria, *i.e.*, wavelength of the absorption maximum and ESR spectrum of the corresponding radical, the chlorinated dibenzo-*p*-dioxins may be readily distinguished from the chlorinated dibenzofurans, the chlorinated diphenylethers, and the chlorinated xanthenes since none of these types of compounds form cation radicals under any of the conditions described in this paper.

### *Acknowledgments*

Helpful discussions with David N. Lincoln are acknowledged with pleasure. The 1,6-dibenzo-*p*-dioxin sample was kindly supplied by J. H. Vinopal and J. E. Casida, Division of Entomology, University of California, Berkeley, Calif.

### *Literature Cited*

1. Tomita, M., Ueda, S., *Chem. Pharm. Bull.* (1964) **12,** 33.
2. Tomita, M., Ueda, S. I., Narisada, M., *Yakugaku Zasshi* (1954) **79,** 186.
3. Shine, H. J., Piette, L., *J. Amer. Chem. Soc.* (1962) **84,** 4798.
4. Shine, H. J., Das, C. F., Small, R. J., *J. Org. Chem.* (1964) **29,** 21.
5. Yang, G. C., Pohland, A. E., unpublished results.
6. Yang, G. C., Pohland, A. E., unpublished results.
7. Tozer, T. N., Tuck, L. D., *J. Chem. Phys.* (1963) **38,** 3035.

RECEIVED October 3, 1972.

# 6

# Photochemistry of Dibenzo-*p*-dioxins

JACK R. PLIMMER and UTE I. KLINGEBIEL

Plant Science Research Division, Agricultural Research Service,
U. S. Department of Agriculture, Beltsville, Md. 20705

DONALD G. CROSBY and ANTHONY S. WONG

Department of Environmental Toxicology, University of California,
Davis, Calif. 95616

*Chlorinated dibenzo-*p*-dioxins are contaminants of phenol-based pesticides and may enter the environment where they are subject to the action of sunlight. Rate measurements showed that 2,3,7,8-tetrachlorodibenzo-*p*-dioxin (TCDD) is more rapidly photolyzed in methanol than octachlorodibenzo-*p*-dioxin. Initially TCDD yields 2,3,7-trichlorodibenzo-*p*-dioxin, and subsequent reductive dechlorination is accompanied by ring fission. Pure dibenzo-*p*-dioxin gave polymeric material and some 2,2′-dihydroxybiphenyl on irradiation. Riboflavin-sensitized photolysis of the potential precursors of dioxins, 2,4-dichlorophenol and 2,4,5-trichlorophenol, in water gave no detectable dioxins. The products identified were chlorinated phenoxyphenols and dihydroxybiphenyls. In contrast, aqueous alkaline solutions of purified pentachlorophenol gave traces of octachlorodibenzo-*p*-dioxin on irradiation.*

Pesticides derived from chlorinated phenols (Table I) are among the most prominent of those currently in worldwide use. Several major herbicides have been applied in large quantities in subtropical locations. California used more than 1,200,000 pounds of 2,4-dichlorophenoxyacetic acid (2,4-D) and its derivatives in 1970 (*1*); Hawaii consumed some 465,000 pounds of pentachlorophenol (PCP) in 1968 (*2*), and the amount of combined butyl esters of 2,4-D and 2,4,5-trichlorophenoxyacetic acid (2,4,5-T) released in one area of Cambodia during two months of 1969 was estimated to exceed 77,000 pounds (*3*).

**Table I. Some Pesticides Based on Polychlorophenols**

| *Common Name (Use)* | *Chemical Name* |
|---|---|
| 2,4-D (herbicide) | 2,4-Dichlorophenoxyacetic acid |
| 2,4-DB (herbicide) | 4-(2′,4′-Dichlorophenoxy) butyric acid |
| Nitrofen (herbicide) | 2,4-Dichlorophenyl 4-nitrophenyl ether |
| Genite (acaricide) | 2,4-Dichlorophenyl benzenesulfonate |
| Zytron (herbicide) | *0*-(2′,4′-Dichlorophenyl) *0*-methyl isopropyl-phosphoramidothioate |
| 2,4,5-T (herbicide) | 2,4,5-Trichlorophenoxyacetic acid |
| Silvex (herbicide) | 2-(2′,4′,5′-Trichlorophenoxy) propionic acid |
| Erbon (herbicide) | 2-(2′,4′,5′-Trichlorophenoxy) ethyl 2,2-dichloropropionate |
| Ronnel (insecticide) | *0,0*-Dimethyl *0*-2,4,5-trichlorophenyl phosphorothioate |
| Gardona (insecticide) | 2-Chloro-1-(2′,4′,5′-trichlorophenyl)-vinyldimethyl phosphate |
| PCP (herbicide, insecticide), wood preservative | Pentachlorophenol |

Variations in the manufacturing process of 2,4,5-trichloro- and pentachlorophenol (but not 2,4-dichlorophenol) have sometimes resulted in contamination of the product by small amounts of heterocyclic impurities (*4, 5*). Of these, the chlorinated dibenzo-*p*-dioxins such as TCDD (2,3,7,8-tetrachlorodibenzo-*p*-dioxin) have received much scientific and public attention because of their real or potential toxicity (*6, 7*). [Chick edema factor, a curious toxicological problem to poultry producers for several years, has been shown to be composed of chlorodibenzo-*p*-dioxins (*8*).]

Phenol-based pesticides may be applied as foliar sprays or otherwise enter the environment where they become subject to the action of sunlight (*7, 8, 9, 10*). Consequently, their contained dioxins also will be exposed to sunlight. The present study determines the extent, conditions, and products of their photodegradation. 2,4,5-T and 2,4-D are degraded in sunlight to a mixture of products in which the parent phenol is prominent (*9, 10*). The possibility that chlorinated phenols might form dioxins by the action of sunlight, as they do by thermal ring closure, also was investigated.

The photolysis of chlorinated aromatic compounds occurs by several processes which follow predictable routes (*13*). They frequently undergo photochemical loss of chlorine by dissociation of the excited molecule to free radicals or, alternatively, through a nucleophilic displacement reaction with a solvent or substrate molecule. Either mechanism is plausible, and the operation of one or the other may be influenced by the reaction medium and the presence of other reagents.

In dilute solution, the structure of the ultraviolet (UV) irradiation products depends on the solvent. In methanol, *o*-chlorobenzoic acid is photoreduced quite rapidly to benzoic acid, whereas a mixture of salicylic and benzoic acids forms slowly in water; however, in benzene the product is 2-phenylbenzoic acid. *m*-Chlorobenzoic acid reacts much more slowly than the *o*-isomer under the same conditions, and a benzoic acid containing *ortho*- and *meta*-chlorine atoms will undergo preferential loss of the ortho substituent (*14, 15*). Although the facile photolytic reduction of halogenated aromatic compounds in methanol and other alcohols has preparative value and has been investigated by several workers (*16, 17*), detailed correlations between position and ease-of-loss of the chlorine substituent and the effect of other substituents on the reaction have not been reported. We must recognize that such photochemical transformations involve an excited state and not the ground-state electronic distribution to which Hammett functions are applicable.

### *The Photolysis of Dioxins*

The biological activity of several halogenated herbicides in water is destroyed by ultraviolet irradiation (*18*). Irradiation seems to be a promising method for decontaminating small quantities of pesticides. The chemical similarity between the chlorinated dioxins and other chlorinted aromatic compounds suggested that if there were parallels in their photochemical behavior, sunlight might destroy dioxins in the environment.

2,3,7,8-Tetrachlorodibenzo-*p*-dioxin (TCDD) (I), an occasional contaminant in 2,4,5-T and other trichlorophenol derivatives, is the most toxic of the commonly-encountered dioxins (*8*) and it received most of our attention. Its low solubility in common solvents and water (*ca.* 2 ppb) limited our experiments since the products were difficult to identify by the conventional techniques of organic chemistry. However, TCDD has an absorption maximum at 307 nm in methanol—well within the solar spectrum observed at the earth's surface and near the region of maximum intensity (310–330 nm) of the UV lamps used in previous experiments (*14, 19*).

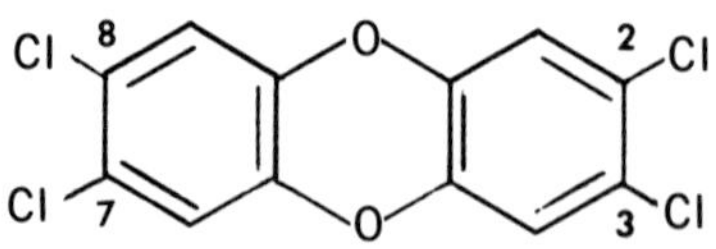


The photolysis rate of several chlorinated dioxins was determined in methanol (*20*) (Figure 1). Solutions of 2,7-dichlorodibenzo-*p*-dioxin (5 mg/liter), TCDD (5 mg/liter), and octachlorodibenzo-*p*-dioxin (2.2 mg/liter) were irradiated with light having an intensity of about 100 $\mu W/cm^2$

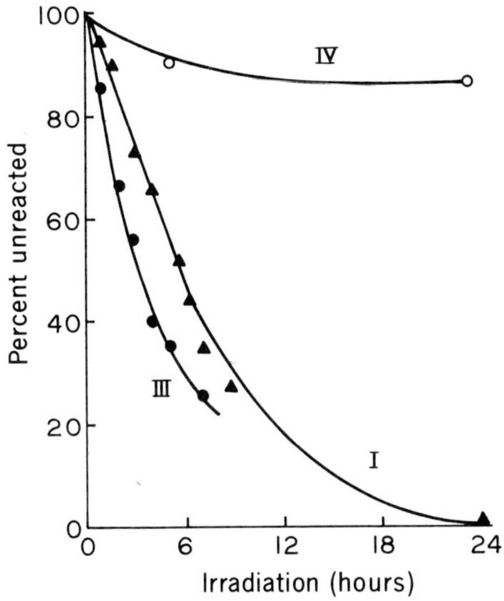


Science

*Figure 1. Photolysis rates of chlorinated dibenzo-*p*-dioxins in methanol under ultraviolet light: 2,7-dichlorodibenzo-*p*-dioxin (III) (5 mg/liter), 2,3,7,8-tetrachlorodibenzo-*p*-dioxin (I) (5 mg/liter), 1,2,3,4,6,7,8,9-octachlorodibenzo-*p*-dioxin (IV) (2.2 mg/liter)* (20), *© 1971 by AAAS*

at the absorption maximum of TCDD (307 nm). Photolysis was performed in sealed borosilicate glass vessels, and the concentration change was determined by gas chromatography (GLC). As expected, reductive dechlorination occurred during irradiation, and as the tetra- or octachloro compounds disappeared, peaks of shorter GLC retention times were observed. TCDD afforded 2,3,7-trichlorodibenzo-*p*-dioxin which was identified by GLC coupled to a mass spectrometer (GC/MS) (Figure 2). A dichlorodibenzo-*p*-dioxin also was formed in low yield.

Octachlorodibenzo-*p*-dioxin was photolyzed much more slowly than TCDD (Figure 1). The rate of dioxin photolysis increased as the number of substituent chlorine atoms decreased. Octachlorodibenzo-*p*-dioxin gave what seemed to be a series of chlorinated dioxins of decreasing chlorine content (*20*).

Similar results were obtained when TCDD in methanol was exposed to natural sunlight in sealed borosilicate glass tubes or beakers (Figure 3). After about 36 hours exposure, a yellow non-volatile gum was obtained as the sole product by evaporation of the solvent. It showed no UV absorption and did not seem to retain the benzenoid chromophore;

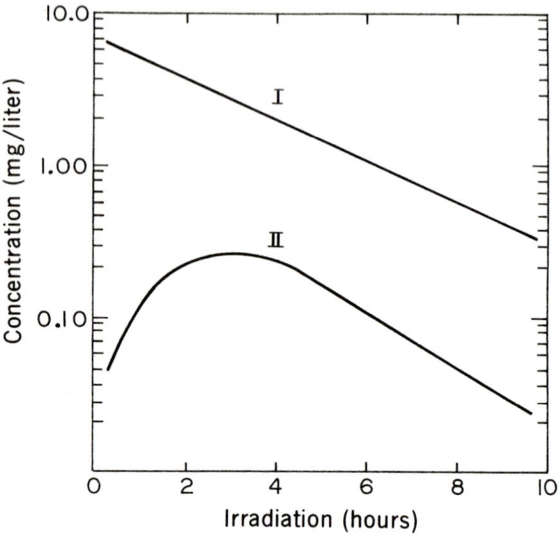


Science

*Figure 2. Photoreduction of 2,3,7,8-tetrachlorodibenzo-*p*-dioxin (I) (2 mg/liter in methanol) as compared with that of the 2,3,7-trichloro-homolog (II)* (20), *© 1971 by AAAS*

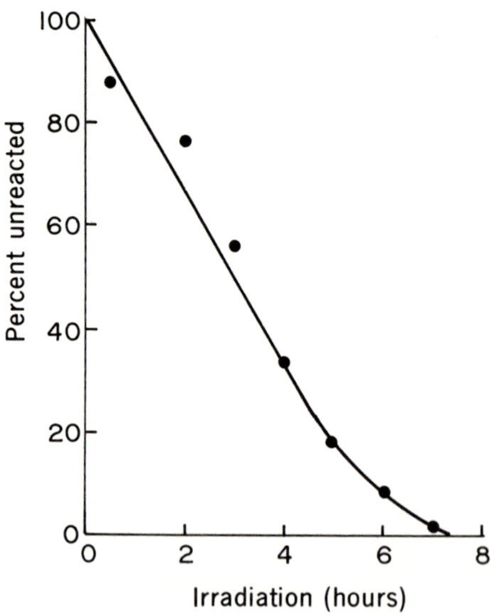


Science

*Figure 3. Photolysis rate of 2,3,7,8-tetrachlorodibenzo-*p*-dioxin (5 mg/liter in methanol) in sunlight* (20), *© 1971 by AAAS*

since unsubstituted dibenzo-*p*-dioxin shows a strong absorption maximum at 290 nm in methanol, it was not present as a reaction product.

It seems that deep-seated cleavage of the dioxin nucleus must accompany dechlorination in methanol. When pure dibenzo-*p*-dioxin (II) was irradiated in cyclohexane solution in a quartz cuvette, it darkened in color, and a precipitate of intractable dark brown material was collected and was insoluble in the common solvents except for methanol.

A more concentrated (1000 ppm) solution of dibenzo-*p*-dioxin in methanol was irradiated for 1.5 hours under a 450-watt lamp fitted with a borosilicate glass filter while nitrogen was bubbled continuously through the solution. Unchanged starting material was recovered to the extent of 85%. The principal photolysis product again was a dark brown insoluble gum similar to that described above. Its mobility on thin layer chromatography (TLC) was very low in the benzene/ethyl acetate (4:1) solvent used to separate the other products.

Under UV light a typical chromatogram showed starting material, a polymeric substance, and two mobile compounds which were isolated by preparative TLC. The more mobile compound has yet to be identified, but the mass spectrum of the least mobile ($R_f$ 0.52) showed the molecular ion at *m/e* 186 and fragment ions at 168 (M-18), 158 (M-28), 157 (M-29), and 139 (M-29-18). The molecular weight of 186 corresponded to either a phenoxyphenol or dihydroxybiphenyl; methylation with diazomethane gave a compound with *m/e* 214 (186 − 2H + two methyl groups) accompanied by fragment ions at *m/e* 185 (M-29), 169 (M-45), 168 (M-46), and 157 (M-57). Since it gave a dimethyl derivative with diazomethane, this product must be a dihydroxybiphenyl, and its identity was confirmed by comparing its chromatographic properties and mass spectrum with those of an authentic specimen of 2,2′-dihydroxybiphenyl (IV) (Figure 4).

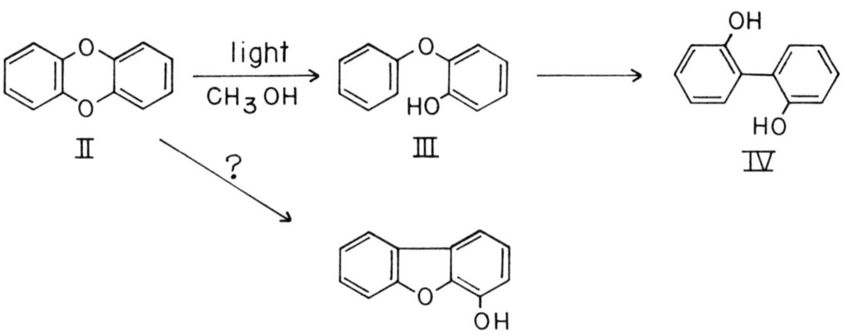


*Figure 4. Irradiation of pure dibenzo-*p*-dioxin*

When the crude reaction mixture was examined by GC/MS, an additional compound with *m/e* 186, which gave a monomethyl ether, was detected in trace amounts. This corresponds to the expected 2-phenoxyphenol (III), the initial product of ring opening (Figure 4).

The photolytic rearrangement of diphenyl ether to *o*- and *p*-phenylphenol has been described by Ogata *et al.* (*21*). The intramolecular reaction proceeded through ether bond fission followed by rearrangement of the radical fragments which recombined to give hydroxylated biphenyls. A trace of phenol was also obtained. The yields were solvent dependent, and it was suggested that formation of a hydrogen bond from the alcohol solvent to the ethereal oxygen atom might aid C—O—C bond fission. It is conceivable that the dioxins yield initially 2-phenoxyphenol which can give symmetrical hydroxybiphenyls by ether bond fission, rearrangement, and radical recombination. Alternatively, dihydroxybenzenoid compounds might be formed and would be readily susceptible to photooxidation, yielding polymeric compounds.

Although the effectiveness of sunlight in destroying TCDD in dilute methanol solution was established conclusively, the insolubility of TCDD limited studies in aqueous systems. Pure crystalline TCDD as a suspension in distilled water seemed to be unchanged by irradiation with a sunlamp. However, if a few drops of a solution of $^{14}$C-TCDD in benzene was added to water and the dispersion was stabilized by a surfactant such as Tween-80, irradiation with a sunlamp was effective in reducing the TCDD content of the system. By contrast, TCDD applied to soil or a solid surface seemed to be extremely resistant to the action of sunlight and decomposed very slowly. A spot of a methanolic solution of TCDD (2.4 ppm) was applied to glass plates coated with a uniform 250-$\mu$m layer of either Norfolk sandy soil or Hagerstown silty clay loam. The dried spot was illuminated for 96 hours with a fluorescent ultraviolet lamp. At the end of this period TCDD could be recovered quantitatively unchanged. The same results were given by dry plates or by plates moistened with water during the experiment. Similarly, there was negligible change in the quantity of TCDD irradiated as a film on a glass plate for up to 14 days (*20*). It seems likely, therefore, that any loss of TCDD from a soil surface would occur by mechanical transfer on dust particles rather than by volatilization or by photodegradation.

### *Photochemical Generation of Dioxins*

Light can effect the coupling of phenols. For example, Joschek and Miller (*22*) found that phenoxyphenols could be produced in the flash photolysis of phenol, but although sought, no dioxin was detected in the reaction products.

The irradiation of aqueous solutions of 2,4-dichlorophenol and 2,4,5-trichlorophenol with UV light at wavelengths above 280 nm gave little reaction. However, in the presence of riboflavin as a sensitizer, the phenols were efficiently consumed (23). Added oxygen seemed to improve reaction yields, perhaps by oxidizing reduced riboflavin.

The reaction products from 2,4-dichlorophenol were tetrachlorophenoxyphenols and tetrachlorodihydroxybiphenyls (Figure 5), as determined from their mass spectra and those of their methyl ethers. 4,6-Dichloro-2-(2′,4′-dichlorophenoxy)phenol (V) was the major phenoxyphenol; the mass spectral fragmentation pattern of *o*-hydroxyphenol ethers is quite characteristic since a hydrogen transfer occurs during the fragmentation (Figure 6). A trace of a trichlorophenoxyphenol also was detected and was formed presumably by the unsensitized reductive loss of chlorine, discussed previously.

Two isomeric tetrachlorodihydroxybiphenyls (*m/e* 322) also occurred in the reaction mixture and were characterized by the formation

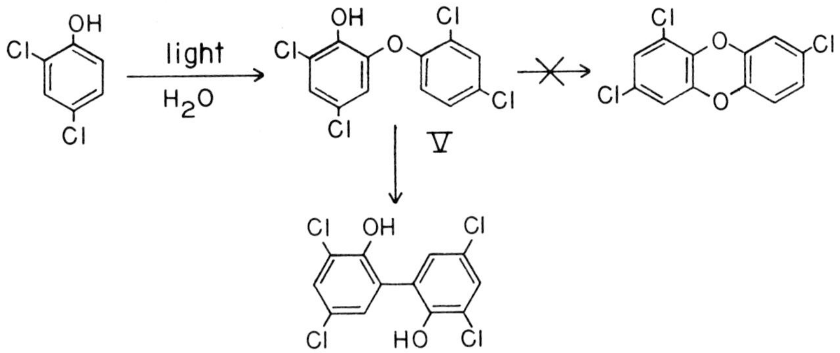


*Figure 5. Irradiation of 2,4-dichlorophenol*

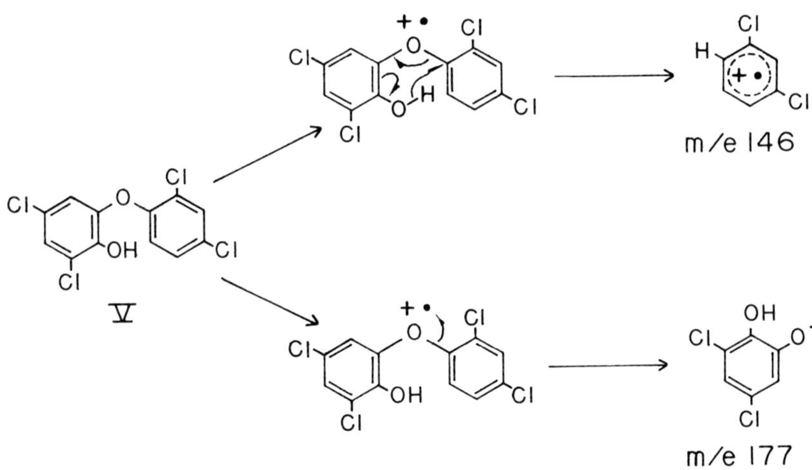


*Figure 6. Fragmentation of 4,6-dichloro-2-(2′,4′-dichlorophenoxy)phenol*

of their dimethyl ethers (*m/e* 350). However, despite extensive destruction of the phenols, their conversion to dimeric products was less than 5%. Although measurements were conducted at several time intervals, at no time could dibenzo-*p*-dioxins be detected in the mixture of products by electron-capture gas chromatography (*20, 23*).

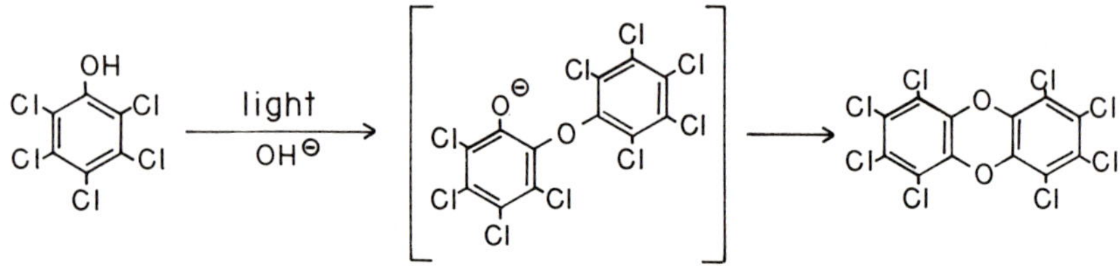


*Figure 7. Photolysis of pentachlorophenol*

By contrast, alkaline aqueous solutions of 2,4-dichlorophenol, 2,4,5-trichlorophenol, and pentachlorophenol rapidly colored when exposed to sunlight. The isolation of several photolysis products predictably derived by hydrolytic and reductive reactions (including tetrachlororesorcinol from pentachlorophenol) suggested strongly that a nucleophilic ionic mechanism is operative (Figure 7). Pure dioxin-free PCP, irradiated as a 1000 ppm solution in aqueous sodium hydroxide with light in the 300–350 nm region, gave a neutral benzene-soluble extract which contained octachlorodibenzo-*p*-dioxin as shown by GLC/MS. Although the yield in repeated experiments was not consistent and the maximum concentration was only 36 ppm in any single irradiation, a smaller amount of an unidentified neutral constituent also was present whose retention time on GLC corresponded to that of a heptachlorodibenzo-*p*-dioxin.

### *Discussion*

It would be difficult to estimate the quantity of TCDD which enters the environment each year. In addition to the common pesticides listed in Table I, other chlorophenols and their derivatives are used widely. For example, large amounts of the disinfectant, hexachlorophene (2,2′-methylenebis(3,4,6-trichlorophenol)), are used in homes, hospitals, and industry, and the Dowcides 2 and B (2,4,5-trichlorophenol and its sodium salt) are industrial microbiocides. More than 50,000,000 lbs of trichlorophenol are made in the United States each year (*24*), and much of it eventually must be dispersed in the environment. The dioxin content seems to be variable but is generally below 0.5 ppm (*25*).

Synthesis of TCDD in sunlight could not be expected to add appreciably to this environmental burden. The formation rate from chlorophenols is strongly concentration-dependent, and its mechanism requires that the two colliding phenol molecules be in anionic form. The rate of trichlorophenol formation from derivatives such as 2,4,5-T is extremely slow, and its dispersal in the environment would be unfavorable to phenol formation at levels required for TCDD synthesis since 2,4,5-T is not converted directly into TCDD. If we consider the rapid rate at which the simpler dioxins undergo photolysis, it seems unlikely that 2,4-D, 2,4,5-T and other chlorophenol-based pesticides—or even the phenols themselves—normally will form the dioxins in the environment.

PCP presents a different picture from that of the lower chlorophenols and their derivatives. The corresponding dioxin shows much more stability to light than does TCDD, enough to permit its prolonged existence at low concentrations in a photoreactor. As a phenol it can directly yield dioxins, a process favored by its normal mode of application as the sodium salt. Although octachlorodibenzo-*p*-dioxin has much lower mammalian toxicity than TCDD (*6*), its formation, properties, and effects demand additional investigation. Technical preparations of PCP are frequently mixtures of tetra- and pentachlorophenols; consequently, hepta- and possibly hexachlorodibenzo-*p*-dioxins might be expected as photolysis products in addition to the octachloro derivative.

TCDD detection has not been reported in the field. Perhaps, this should not be surprising, considering the serious difficulty of its analysis, its photolytic breakdown, and the fact that a normal application (2 lbs/acre) of 2,4,5-T, for example, usually would represent the dispersal of no more than 450 $\mu$g of the dioxin on each treated acre, assuming a dioxin content of $<$0.5 ppm. Likewise, there seem to be no reports of the decay of highly dispersed dioxins on leaf surfaces, and it must be recognized that the experiments we have reported here were conducted only under the artificial conditions of the laboratory. Nevertheless, the photochemistry of the dioxins should provide a worthwhile reminder that sunlight can act on man-made chemicals in the environment to form products which may be more toxic than the original to some life form and that a compound shown to be highly toxic under rigidly standardized conditions does not necessarily present the same hazard when the toxicant is exposed to the forces of nature.

## *Literature Cited*

1. California Department of Agriculture, "Pesticide Use Report 1970," Sacramento, California, 1971.
2. Kimura, H. S., Hurov, H. R., "Evaluation of Pesticide Problems in Hawaii," Appendix, p. 25, Hawaii Department of Agriculture, Honolulu, 1969.
3. Whiteside, T., "Defoliation," p. 129, Ballantine, New York, 1970.
4. Kimmig, J., Schulz, K. H., *Dermatologica* (1957) **195,** 540.
5. Bauer, H., Schulz, K. H., *Arch. Pathol. Gewerbehyg.* (1961) 583.

6. Panel on Herbicides, President's Science Advisory Committee, "Report on 2,4,5-T," Executive Office of the President, Washington, D. C. (1971).
7. Emerson, J. L., Thompson, D. J., Strebing, R. J., Gerbig, C. G., Robinson, V. B., *Fd. Cosmet. Toxicol.* (1971) **9**, 395.
8. Higginbotham, G. R., Huang, A., Firestone, D., Verrett, J., Ress, J., Campbell, A. D., *Nature* (1968) **220**, 702.
9. Crosby, D. G., Tutass, H. O., *J. Agr. Food Chem.* (1966) **14**, 596.
10. Crosby, D. G., Wong, A. S., *161st Nat. Meet. Amer. Chem. Soc.*, Los Angeles (April 1971).
11. Crosby, D. G., Hamadmad, N., *J. Agr. Food Chem.* (1971) **19**, 1171.
12. Kuwahara, M., Kato, N., Munakata, K., *Agr. Biol. Chem. (Tokyo)* (1966) **30**, 232, 239.
13. Plimmer, J. R., *Residue Rev.* (1971) **33**, 47.
14. Plimmer, J. R., Hummer, B. E., *155th Nat. Meet. Amer. Chem. Soc.*, San Francisco (April 1968).
15. Crosby, D. G., Leitis, E., *J. Agr. Food Chem.* (1969) **17**, 1033.
16. Pinhey, J. T., Rigby, R. D. G., *Tetrahedron Lett.* **1970**, 1267.
17. Wolf, W., Kharasch, N., *J. Org. Chem.* (1965) **30**, 2493.
18. Kearney, P. C., Woolson, E. A., Plimmer, J. R., Isensee, A. R., *Residue Rev.* (1969) **29**, 137.
19. Crosby, D. G., *Residue Rev.* (1969) **25**, 1.
20. Crosby, D. G., Wong, A. S., Plimmer, J. R., Woolson, E. A., *Science* (1971) **173**, 748.
21. Ogata, Y., Tagaki, K., Ishino, I., *Tetrahedron* (1970) **26**, 2703.
22. Joschek, H. I., Miller, S. I., *J. Amer. Chem. Soc.* (1971) **88**, 407.
23. Plimmer, J. R., Klingebiel, U. I., *Science* (1971) **174**, 407.
24. United States Tariff Commission, "Synthetic Organic Chemicals," p. 52, TC Publication 295, Washington, D. C., 1969.
25. Woolson, E. A., Thomas, R. F., Ensor, P. D. J., *J. Agr. Food Chem.* (1972) **20**, 351.

RECEIVED February 23, 1972.

7

# Toxicology of Chlorinated Dibenzo-*p*-dioxins

B. A. SCHWETZ, J. M. NORRIS, G. L. SPARSCHU, V. K. ROWE, and P. J. GEHRING

Chemical Biology Research, The Dow Chemical Co., Midland, Mich. 48640

J. L. EMERSON and C. G. GERBIG

Human Health Research and Development Center, The Dow Chemical Co., Zionsville, Ind., 46077

*Severe toxicological responses have been attributed to certain chlorodibenzo-*p*-dioxins. Therefore, the acute and subacute toxicities of several of these compounds have been determined and these results provide information for evaluating potential health hazards. 2,3,7,8-Tetrachlorodibenzo-*p*-dioxin (2,3,7,8-TCDD) was the most toxic chlorodibenzo-*p*-dioxin studied, having an $LD_{50}$ in the μg/kg range. Hexachlorodibenzo-*p*-dioxin (HCDD) was less toxic than 2,3,7,8-TCDD but more toxic than either 2,7-dichlorodibenzo-*p*-dioxin (2,7-DCDD) or octachlorodibenzo-*p*-dioxin (OCDD). Both 2,3,7,8-TCDD and HCDD were acnegenic, highly embryotoxic, and positive for the chick edema factor. 2,7-DCDD and OCDD were not chloracnegenic, caused little or no embryotoxicity, and were low in acute oral lethality; OCDD was negative for the chick edema factor.*

Severe toxicological responses have been associated with certain chlorodibenzodioxins. One of these responses is chloracne, a folliculosis first associated with skin contamination by chlorohydrocarbons in 1899 (*3*). Serious outbreaks of chloracne-like lesions associated with runaway reactions in the production of 2,4,5-trichlorophenol occurred in Germany in the early 1950's (*5*). 2,4,5-Trichlorophenol itself does not cause acne (*8*), but the contaminants which may be formed in the uncontrolled production of 2,4,5-trichlorophenol are extremely potent acnegens (*5*). 2,3,7,8-Tetrachlorodibenzo-*p*-dioxin and tri- and tetrachlorodibenzofuran were isolated from the contaminants formed in 2,4,5-trichlorophenol production and were demonstrated to be strongly positive acnegens when applied to rabbit ears (*8*). Using the rabbit ear test, the acnegenic potency of 2,3,7,8-tetrachlorodibenzo-*p*-dioxin (2,3,7,8-TCDD) was confirmed in 1962 (*6*). In addition, 2,3,7,8-TCDD is extremely toxic in the chick embryo assay (*4*) and is highly embryotoxic in rats (*12*). Another chlorodibenzodioxin, hexachlorodibenzo-*p*-dioxin (HCDD), is known to be positive for the chick edema factor, a condition characterized by hydropericardium, ascites, and anasarca (*1, 4*).

## *Experimental*

**Materials.** The chlorodibenzodioxin samples used in these studies are identified and described in Table I. Studies were limited in some cases by availability of pure samples.

**Acute Lethality.** Samples of 2,7-dichlorodibenzo-*p*-dioxin, 2,3,7,8-tetrachlorodibenzo-*p*-dioxin, hexachlorodibenzo-*p*-dioxin, and octachlorodibenzo-*p*-dioxin were evaluated for acute oral lethality in the following animals.

| | | *Test Material* | | | |
|---|---|---|---|---|---|
| *Test Animal* | *Strain* | *2,7-DCDD* | *2,3,7,8-TCDD* | *HCDD* | *OCDD* |
| Rat | Sprague-Dawley | X | | X | X |
| Rat | Sherman (Spartan) | | X | | |
| Mouse | Swiss Webster | X | X | | X |
| Rabbit | New Zealand albino | | X | | |
| Guinea pig | Hartley | | X | | |
| Dog | Beagle | | X | | |

Test materials were administered as suspensions in corn oil or as corn oil:acetone (9:1) solutions in single doses by gavage. The animals were deprived of feed for 16 hours before dosing. After dosing they were observed for signs of toxicity including body weight changes for two to eight weeks.

Lethality of 2,3,7,8-TCDD *via* skin absorption was tested on rabbits of mixed sexes with doses of 31.6, 63, 126, 252, and 500 $\mu$g/kg body weight. The compound was applied as a 0.01% solution in acetone to the abdominal skin which had been shaven. After the acetone evaporated, the trunk of each rabbit was wrapped in cotton to prevent ingestion. The rabbits were housed in individual holding cages and were observed for signs of toxicity including body weight changes for three weeks.

Parenteral lethality was determined by injecting rabbits of mixed sexes intraperitoneally with 31.6, 63, 126, 252, and 500 $\mu$g/kg of 2,3,7,8-TCDD as a 0.01% corn oil suspension; control rabbits were injected with corn oil. The rabbits were housed in individual holding cages and were observed for signs of toxicity for four weeks. The $LD_{50}$'s were calculated by the Weil modification of the Thompson method (*14, 15*) or by the Litchfield and Wilcoxon method (*9*). The acute lethality studies were terminated when it was evident that the survivors were not showing signs of toxicity.

**Table I. Purity of Samples Used in the Toxicology Studies**

| *Sample Identification* [a] | *Purity* [b] | *Tests* [c] |
|---|---|---|
| 1. 2,7-Dichlorodibenzo-*p*-dioxin | | |
| a. No. 104, shelf 142 | 99.8% | 1, 2, 3 |
| b. AR-570 [d] | | 3 |
| c. 340-2-13A | 99.6% | 1, 2, 3 |
| d. 340-2-69A | >99% | 4 |
| 2. 2,3,7,8-Tetrachlorodibenzo-*p*-dioxin | | |
| a. Caustic insoluble isolate 1965 | 96.4% | 1 |
| b. 851-142-24 | 98% | 1 |
| c. Skelly 11/11/64 | 91% | 1, 3, 5 |
| d. 340-2-54B | >99% | 1, 2, 3, 5 |
| 3. 1,2,3,4-Tetrachlorodibenzo-*p*-dioxin | | |
| a. FDA-F990 | 98.5% | 3 |
| 4. Hexachlorodibenzo-*p*-dioxin | | |
| a. 252-44-12B-AL22 | 65:35, 2 isomers | 1, 3 |
| b. 252-44-12B-AL11 | 99%, 65:35, 2 isomers | 3 |
| c. 340-2-82A | >99%, 89:11, 2 isomers | 1, 2, 3, 4, 5 |
| d. FDA-F911 | 95.1%, 3 isomers | 3 |
| 5. Octachlorodibenzo-*p*-dioxin | | |
| a. 251-1-142A | 98% | 1, 2, 3 |
| b. 340-2-29A | 94% | 1, 3 |
| c. AR-570 [e] | | 3 |
| d. 340-2-57A | 98.86% | 1, 3, 4, 5 |

[a] All samples are from The Dow Chemical Co. except 3a and 4d, which are from the Food and Drug Administration.
[b] Based on gas-liquid chromatographic (GLC) or GLC-mass spectrophotometric analysis.
[c] Test identification: 1 = $LD_{50}$, 2 = eye irritation, 3 = chlorance, 4 = teratogenicity, 5 = chick edema.
[d] Photolysis product of 1a.
[e] Photolysis product of 5a.

**Eye Irritation.** Rabbit eyes were examined prior to experiments and found to be free of defects or irritation. Approximately 2 mg of 2,7-DCDD, 2,3,7,8-TCDD, HCDD, or OCDD were instilled in the conjunctival sac of one eye; the contralateral eye served as a control. The eyes were examined at various times after treatment for conjunctival redness and chemosis, iritis, and corneal injury. Responses were catagorized according to intensity.

**Rabbit Ear Bioassay for Acnegenic Activity.** Acnegenic activity of 2-7-DCDD, 1,2,3,4-TCDD, 2,3,7,8-TCDD, HCDD, and OCDD was tested by applying 0.1 ml of either a solvent solution or the supernatant of a solvent suspension of each compound to the inner surface of the rabbit's ears five days a week for four weeks. The ears were examined weekly for signs of chloracne, inflammation, and hyperkeratosis. The responses were divided into five categories: (1) none, (2) very slight, (3) slight, (4) moderate, and (5) severe.

Responses in the first three categories include no response to mild irritation, increased ear thickness, slight enlargement of the follicular aperture, slight exfoliation, and slight crust formation. These responses alone are not considered indicative of chloracnegenic activity. Categories 4 and 5 are indicative of acnegenic response and are characterized by comedo formation, increased ear thickness, and hyperkeratosis.

**Teratology.** Pregnant adult Sprague-Dawley (Spartan strain) female rats weighing approximately 250 grams were used to study teratogenicity of the chlorinated dibenzo-*p*-dioxins. The day sperm were first present in a vaginal smear was considered day zero of pregnancy. The animals were housed individually in wire-bottom cages in a room controlled for temperature, humidity, light cycle, and noise. Commercial laboratory rat chow and water were provided with choice.

Corn oil:acetone (9:1) solutions, with varying amounts of test material were given in 2.5 ml/kg dosages by gavage. Dosages were calculated using daily body weights. Rats were treated with 100 mg of 2,7-DCDD/kg/day, 0.1, 1.0, 10, or 100 $\mu$g HCDD/kg/day and 100 or 500 mg OCDD/kg/day on days 6 through 15 of gestation. Control rats received 2.5 ml/kg of corn oil:acetone (9:1) orally. All rats were observed daily throughout pregnancy and were weighed on days 6, 13, and 21 of gestation.

Pregnant females were sacrificed by carbon dioxide anesthesia on day 21 of gestation; the uterine horns were exteriorized through a midline incision in the abdominal wall, and the number and position of live, dead, and resorbed fetuses were noted. After being weighed and sexed, the fetuses were examined for external anomalies; the crown-rump length was measured with a vernier caliper. Half of each litter was preserved in Bouin's solution and later examined for soft tissue anomalies (*16*); the other half was preserved in alcohol, cleared and stained with Alizarin red-S, and examined for skeletal abnormalities (*2*).

A 2 $\times$ 2 contingency table was used to evaluate the frequency of anomalies and resorptions within the fetal population and between litters. Body weight and body measurements were statistically analyzed by an Analysis of Variance and Tukey's test (*13*). In all cases, the level of significance was $P < 0.05$.

**Chick Bioassay for Chick Edema Factor.** The bioassay for chick edema factor was conducted according to the "Official Methods of Analysis," 10th Ed., Sections 26.087–26.091, Association of Official Agricultural Chemists. Three-day-old white leghorn, single-comb cockerels were used. 2,3,7,8-TCDD, HCDD, and OCDD were the compounds studied. The diet used in the study was formulated specifically for conducting the chick edema bioassay ("Nutritional Biochemicals," International Chemical and Nuclear Corp., Cleveland, Ohio). Body weights were recorded twice weekly for the oral intubation studies and at the start and termination of the dietary study. The chicks were observed daily for signs of toxicity, and food consumption was recorded weekly. After 20 or 21 days of treatment, all chickens were sacrified by cervical dislocation and examined for gross lesions. The amount of pericardial and peritoneal fluid was measured, and all gross lesions were recorded. If the calculated "*t*" was greater than +1.3, the mean logarithm (100 $\times$ ml pericardial fluid) was greater than 1.1461 for the chicks receiving the test compound, and the mean logarithm of the negative control was less than 1.1460, the compound was considered positive for chick edema.

**Pathology.** Toxicology studies were not designed to study the pathological changes associated with chlorodibenzodioxin administration, but in some cases, gross pathological and histopathological examinations were performed. For microscopic examination, tissues were fixed in 10% buffered formalin and were stained with hematoxylin and eosin. Sections of fetuses of control dams and dams treated with 100 mg 2,7-DCDD/kg/day were stained with hematoxylin and eosin, hematoxylin-phloxine-saffron, Masson's trichrome stain, and Mallory's phosphotungstic acid hematoxylin stain.

## *Results*

**Acute Lethality.** The lethality of 2,3,7,8-TCDD is presented in Table II. The data reveal that the single oral $LD_{50}$ ranges from 0.0006 mg/kg in male quinea pigs to 0.115 mg/kg in rabbits of mixed sex. Data on rats and guinea pigs indicate that males are more sensitive than females; lethality is essentially the same following intraperitoneal, oral, or skin administration for rabbits. Limited data show that dogs are less sensitive to 2,3,7,8-TCDD than rabbits. For female and male mice, single oral doses ranging from 0.001 to 0.130 mg/kg produced a few sporadic deaths without any definitive dose-response relationship; therefore the data are not presented in the table.

Limited lethality data are available for 2,7-DCDD, HCDD, and OCDD. HCDD (sample c) killed 1 of 2 and 0 of 2 male rats given oral doses of 100 and 10 mg/kg, respectively. No deaths occurred in four male mice given 2.0 grams/kg of 2,7-DCDD (sample a or b) orally or in two female rats given 1 gram/kg (sample a). For OCDD, oral doses of 1 gram/kg (sample d) to five female rats did not cause death; in four male mice, doses of 4 grams/kg also did not cause death. No signs of toxicity were observed in animals treated with either 2,7-DCDD or OCDD. The only sign of toxicity among animals treated with HCDD was loss of body weight.

While all species lost body weight following treatment with 2,3,7,8-TCDD, other signs of toxicity were species dependent. Ascites was seen in mice. Anorexia, dehydration, depression, emaciation, intestinal hemorrhage, and alopecia were seen in dogs. Certain rabbits treated intraperitoneally with 2,3,7,8-TCDD developed skin lesions typical of those associated with acnegens.

**Rabbit Eye Irritation.** Instillation of the chlorodibenzodioxins into the conjunctival sac caused slight, transient pain and conjunctival inflammation, initially. Treatment with 2,3,7,8-TCDD was associated with delayed conjunctival chemosis 13–22 days later. By day 27, the chemosis had subsided, but the rim of the eyelid was thickened and encrusted.

**Table II. Lethality of 2,3,7,8-**

| *Species, Sex* [b] | *Administration* [b] | *Time of Death, Days Postadministration* |
|---|---|---|
| Rat, male | Oral | 9–27 |
| Rat, female | Oral | 13–43 |
| Guinea pig, male | Oral | 5–34 |
| Guinea pig, female | Oral | 9–42 |
| Rabbit, mixed | Oral | 6–39 |
| | Skin | 12–22 |
| | Intraperitoneal | 6–23 |
| Dogs, male | Oral | 9–15 |
| Dogs, female | Oral | — |

[a] Responses to individual doses are given in those cases in which an $LD_{50}$ could not be calculated. The $LD_{50}$ for oral administration to rabbits was calculated using the method of Litchfield and Wilcoxon (*9*); the remaining values were calculated using the Weil modification of the method of Thompson (*14, 15*).

In rabbits treated with HCDD, the rim of the eyelid was encrusted 27 days after treatment. Neither corneal injury nor iritis was observed in any of the animals following instillation of the chlorodibenzodioxins in the conjunctival sac.

**Acnegenic Response.** Both 2,3,7,8-TCDD and HCDD produced chloracne in the rabbit ear bioassay as indicated by the formation of comedones. Solutions of 2,3,7,8-TCDD (sample c) in benzene ranging in concentration from 0.04 μg/ml to 400 μg/ml produced a positive response with severity increasing with concentration. A negative response was obtained with a solution of 0.004 μg/ml. In contrast, a chloroform solution of 1,2,3,4-TCDD, 50 μg/ml, did not produce a positive response. With HCDD (samples a, b, c, and d), a response was produced by solutions of 10 to 50 μg/ml in chloroform and dimethoxyethane. Chloroform extracts from 10% suspensions of 2,7-DCDD or OCDD were negative, indicating that these have a low order or possibly no acnegenic activity.

**Teratogenicity.** The effects of chlorodibenzodioxins on maternal and fetal body measurements, incidence of fetal resorptions, and anomalies are given in Tables III and IV.

**Tetrachlorodibenzo-*p*-dioxin**[a]

| *$LD_{50}$, mg/kg* | *Dose, mg/kg* | *Number Deaths/ Number Treated* |
|---|---|---|
| 0.022 | 0.008 | 0/5 |
| | 0.016 | 0/5 |
| | 0.032 | 10/10 |
| | 0.063 | 5/5 |
| 0.045 (0.030–0.066) | | |
| 0.0006 (0.0004–0.0009) | | |
| 0.0021 (0.0015–0.0030) | | |
| 0.115 (0.038–0.345) | | |
| 0.275 (0.142–0.531) | | |
| — | 0.032 | 0/5 |
| | 0.063 | 2/5 |
| | 0.126 | 2/5 |
| | 0.252 | 2/5 |
| | 0.500 | 3/5 |
| | 0.30 | 0/2 |
| | 3.00 | 2/2 |
| | 0.03 | 0/2 |
| | 0.10 | 0/2 |

[b] All samples used are from 2c in Table I except guinea pig, female, which is from 2d in Table I.

2,7-DCDD. Rats treated with 100 mg/kg/day on days 6 through 15 of gestation gained slightly more weight during pregnancy than controls but showed no toxicity. There was no effect on fetal body measurements, or incidence of resorptions, or gross, soft tissue, or skeletal anomalies.

HCDD. Administration of 0.1–100 μg HCDD/kg/day was associated with a dose-related decrease in maternal weight-gain during gestation. Gross necropsy examination at the time of cesarean section revealed evidence of maternal toxicity only among dams receiving 100 μg/kg/day (pale, friable liver 3/20 dams; serous atrophy of fat, 1/20 dams).

Treatment with 10 or 100 μg HCDD/kg/day was highly lethal to fetuses during late gestation. While the incidence of early resorptions was not increased at any dose level of HCDD (5–7% in the treated *vs.* 7% in the controls), there was a significant increase in late resorptions (0% at 0.1 μg/kg/day to 79% at 100 μg/kg/day). The weight and length of surviving fetuses were significantly decreased.

A significant increase in the incidence of fetal soft-tissue and skeletal anomalies was seen following treatment of pregnant rats with HCDD at the 100 μg/kg/day dose level. The incidence of cleft palate, subcutaneous edema, vertebrae with split or unfused centra, and split sternebrae was significantly greater than among control litters or the control fetal popu-

**Table III. Effect of Treatment with Chlorinated Measurements and the**

| *Test Compound (Sample)* [a] | *No. of Litters* | *Maternal Weight Gain* [b] *on Gestation Days* 6–13 | 13–21 | 6–21 |
|---|---|---|---|---|
| Control | 30 | 36 ± 2 | 101 ± 6 | 137 ± 8 |
| 2,7-Dichlorodibenzo-*p*-dioxin, (d), mg/kg/day | | | | |
| 100.0 | 7 | 31 ± 1 | 122 ± 4 | 152 ± 5 |
| Hexachlorodibenzo-*p*-dioxin, (c), μg/kg/day | | | | |
| 0.1 | 19 | 28 ± 2 | 102 ± 5 | 130 ± 5 |
| 1.0 | 19 | 27 ± 3 | 99 ± 5 | 126 ± 6 |
| 10.0 | 18 | 22 ± 3 [f] | 97 ± 5 | 119 ± 6 |
| 100.0 | 19 | 6 ± 2 [f] | 13 ± 7 [f] | 19 ± 9 [f] |
| Octachlorodibenzo-*p*-dioxin, (d), mg/kg/day | | | | |
| 100.0 | 12 | 32 ± 2 | 100 ± 8 | 131 ± 7 |
| 500.0 | 17 | 35 ± 3 | 115 ± 4 | 150 ± 5 |

[a] Sample identified in Table I; administered on days 6–15 of gestation as a corn oil: acetone (9:1) solution.
[b] Mean ± S.E.
[c] Mean of litter means ± S.E.

**Table IV. Effect of Treatment with Hexachlorodibenzo-**

| | | *μg/kg/day on Days* 0 | | 0.1 | |
|---|---|---|---|---|---|
| Soft Tissue Anomalies | | | | | |
| Cleft Palate | P [a] | 0 | (0/156) | 1 | (1/104) |
| | L [b] | 0 | (0/28) | 5 | (1/19) |
| Dilated Renal Pelvis | P | 0.6 | (1/156) | 0 | (0/104) |
| | L | 4 | (1/28) | 0 | (0/19) |
| Subcutaneous Edema | P | 5 | (8/156) | 6 | (6/104) |
| | L | 21 | (6/28) | 32 | (6/ 19) |
| Skeletal Anomalies | | | | | |
| Split Vertebral | P | 6 | (9/158) | 2 | (2/103) |
| Centra | L | 19 | (5/27) | 5 | (1/19) |
| Split Sternebrae | P | 0.6 | (1/158) | 1 | (1/103) |
| | L | 4 | (1/27) | 5 | (1/19) |
| Delayed Ossification | P | 11 | (18/158) | 28 | (29/103) [c] |
| of Sternebrae | L | 44 | (12/27) | 74 | (14/19) |

[a] Incidence among fetal population; % (number of affected fetuses/number examined).

**Dibenzo-*p*-dioxins on Maternal and Fetal Body Incidence of Fetal Resorptions**

| Fetal Body Weight,[c] g | Fetal Crown-Rump[c] Length, mm | % Fetal Resorptions: Population [d] | % Fetal Resorptions: Litter [e] |
|---|---|---|---|
| 5.68 ± 0.05 | 44.5 ± 0.1 | 7 (22/337) | 47 (14/30) |
| 5.80 ± 0.09 | 44.2 ± 0.2 | 6 (5/86) | 57 (4/7) |
| 5.73 ± 0.04 | 43.8 ± 0.1 | 5 (10/217) | 47 (9/19) |
| 5.93 ± 0.16 | 45.7 ± 0.5 | 9 (20/218) | 74 (14/19) |
| 5.12 ± 0.05 [f] | 42.6 ± 0.2 [f] | 25 [f] (57/229) | 94 [f] (17/18) |
| 3.65 ± 0.28 [f] | 35.2 ± 0.7 [f] | 85 [f] (194/227) | 100 [f] (19/19) |
| 5.73 ± 0.09 | 43.6 ± 0.4 | 8 (11/131) | 42 (5/12) |
| 5.69 ± 0.05 | 44.5 ± 0.2 | 5 (9/199) | 41 (7/17) |

[d] % (number resorptions/number implantations).
[e] % (number litters with at least one resortion/number litters).
[f] Significantly different from control by an Analysis of Variance and Tukey's test (measurements) or the 2 × 2 contingency table (resorptions), P<0.05.

***p*-dioxin on the Incidence of Fetal Anomalies**

*6–15 of Gestation*

| 1.0 | 10 | 100 |
|---|---|---|
| 0 (0/99) | 0 (0/86) | 47 (8/17) [c] |
| 0 (0/19) | 0 (0/18) | 73 (8/11) [c] |
| 2 (2/99) | 6 (5/86) [c] | 12 (2/17) [c] |
| 5 (1/19) | 17 (3/18) | 18 (2/11) |
| 55 (54/99) [c] | 100 (86/86) [c] | 100 (17/17) [c] |
| 100 (19/19) [c] | 100 (18/18) [c] | 100 (11/11) [c] |
| 1 (1/99) | 7 (6/86) | 31 (5/16) [c] |
| 6 (1/18) | 29 (5/17) | 56 (5/9) [c] |
| 2 (2/99) | 2 (2/86) | 31 (5/16) [c] |
| 11 (2/18) | 12 (2/17) | 56 (5/9) [c] |
| 12 (12/99) | 34 (29/86) [c] | 56 (9/16) [c] |
| 50 (9/18) | 71 (12/17) | 56 (5/9) |

[b] Incidence among litters; % (number of affected litters/number examined).
[c] Significantly different from control by 2 × 2 contingency table, P<0.05.

lation. Among dams treated with 1 or 10 μg/kg/day, only subcutaneous edema occurred at a significantly greater incidence than in the control litters or fetal population. Treatment with 0.1 μg/kg/day of HCDD did not increase fetal anomalies among the litters or the fetal population. The incidence of delayed ossification of sternebrae was significantly increased among the fetal population but not among litters.

OCDD. Signs of maternal toxicity were not observed in rats given 100 or 500 mg/kg/day OCDD. Examination of the fetuses did not reveal changes in fetal body measurements, incidence of fetal resorptions, or incidence of any fetal anomaly among litters or the fetal population. At 500 mg/kg/day, the incidence of subcutaneous edema was significantly increased among the fetal population (23/100 compared with 8/156 in controls) but not among litters (9/18 compared with 6/28 in controls).

**Chick Edema Bioassay.** Chick edema was produced in groups of birds treated with 1 and 10 μg/kg/day of 2,3,7,8-TCDD and 10 and 100 μg/kg/day of HCDD (Table V). The mean logarithm for pericardial fluid volume of the negative control groups was greater than 1.1460 and

**Table V. Results of Chick Edema Bioassay: Body Weight, Calculations of Chicks**

| *Treatment (Sample)* [a] | *n* | *Body Weight g* [b] *Day 0* | *Body Weight g* [b] *Day 21* | *Food Consumption, g* [c] |
|---|---|---|---|---|
| 2,3,7,8-Tetrachlorodibenzo-*p*-dioxin (d) [d], *μg/kg* | | | | |
| 0 | 10 | 46 ± 1 | 199 ± 5 | 17.4 |
| 0.01 | 10 | 44 ± 1 | 196 ± 7 | 16.7 |
| 0.10 | 10 | 45 ± 1 | 203 ± 7 | 17.2 |
| 1.0 [f] | 2 | 42 ± 1 | 196 ± 24 | 33.7 |
| 10.0 [g] | 9 | 42 ± 1 | no survivors | 11.2 |
| Hexachlorodibenzo-*p*-dioxin (c) [d], μg/kg | | | | |
| 0 | 9 | 38 ± 1 | 194 ± 5 | 17.4 |
| 0.1 | 10 | 42 ± 1 | 197 ± 4 | 17.8 |
| 0.0 | 10 | 43 ± 1 | 196 ± 6 | 18.2 |
| 10.0 [h] | 9 | 38 ± 1 | 187 ± 7 | 16.7 |
| 100.0 [i] | 10 | 36 ± 1 | no survivors | 13.3 |
| Octachlorodibenzo-*p*-dioxin (d) [e], % of diet | | (Day 20) | | |
| 0 | 12 | 45 ± 1 | 141 ± 8 | 13.3 |
| 0.1 | 11 | 45 ± 1 | 124 ± 5 | 9.5 |
| 0.5 | 11 | 43 ± 1 | 196 ± 10 | 10.9 |

[a] Sample identified in Table I.
[b] Mean ± SE.
[c] Grams/chick/day.
[d] Administered orally as a corn oil: acetone solution.
[e] Fed in the diet 0.1% ≅ 100 mg/kg, 0.5% ≅ 500 mg/kg).

could negate the results if the guidelines for interpreting chick edema bioassay studies were rigidly followed. However, since the volume of pericardial fluid was markedly increased by the treatments indicated above, the treatments were considered to be positive for the production of chick edema. A positive response was not observed in chicks maintained on a diet containing 0.5% OCDD.

Severe dyspnea, subcutaneous edema, and distended abdomens were observed in some birds receiving 1 or 10 μg 2,3,7,8-TCDD/kg/day. Dyspnea and mucus accumulation in the mouth prior to death were observed in birds receiving 100 μg 2,3,7,8-TCDD/kg/day. No overt clinical signs were observed in birds receiving OCDD.

The gross lesions seen in chicks treated with chlorodibenzodioxins are summarized in Table VI. The most consistent gross lesions were increased pericardial and peritoneal fluid, subcutaneous and pulmonary edema, hepatomegaly, and a mottled appearance of the liver.

Histopathologic examination of tissues of selected birds from the 2,3,7,8-TCDD (1 and 10 μg) and HCDD (10 and 100 μg) groups revealed

**Food Consumption, and Pericardial Fluid Volume Treated with Chlorodioxins**

| *Pericardial Fluid Volume* | | | *Positive for Chick Edema Factor Based on* | |
|---|---|---|---|---|
| *ml ± SE* | *Mean Log 100 ×* | *Calculated t* | *Calculations* | *Gross Lesions* |
| 0.16 ± 0.02 | 1.1717 | — | — | — |
| 0.14 ± 0.02 | 1.1181 | −0.74 | no | no |
| 0.19 ± 0.01 | 1.2688 | +1.69 | no | no |
| 2.34 ± 0.08 | 2.3680 | +21.7 | no | yes |
| 1.29 ± 0.62 | 1.5661 | +1.47 | no | yes |
| 0.15 ± 0.01 | 1.1771 | — | — | — |
| 0.11 ± 0.02 | 0.9978 | −1.93 | no | no |
| 0.09 ± 0.01 | 0.9387 | −4.57 | no | no |
| 0.81 ± 0.01 | 1.7294 | +3.82 | no | yes |
| 0.62 ± 0.24 | 1.5650 | +2.72 | no | yes |
| 0.08 ± 0.01 | 0.8053 | — | — | — |
| 0.06 ± 0.01 | 0.7889 | −0.21 | no | no |
| 0.09 ± 0.01 | 0.9002 | +0.91 | no | no |

[f] Animals died on days, 9, 11, 11, 14, 15, 17, 18 and 19 of treatment.
[g] Animals died on days 3, 4, 4, 4, 5, 8, 8, 9, 12 and 15 of treatment.
[h] One animal died on day 19 of treatment.
[i] Animals died on days, 5, 5, 6, 7, 8, 10, 11, 11, 15 and 17 of treatment.

Table VI. Results of Chick Edema Bioassay: Summary of

| *Treatment (Sample)* [a] | *Mortality* | *Pericardial Fluid >0.2 ml* | *Peritoneal Fluid* | *Subcutaneous Edema* |
|---|---|---|---|---|
| 2,3,7,8-Tetrachlorodibenzo-*p*-dioxin (d) [c], μg/kg | | | | |
| 0 | 0/10 [b] | 2/10 | 0/10 | 0/10 |
| 0.01 | 0/10 | 1/10 | 0/10 | 0/10 |
| 0.10 | 0/10 | 2/10 | 0/10 | 0/10 |
| 1.0 | 8/10 | 10/10 | 9/10 | 9/10 |
| 10.0 | 10/10 | 5/10 | 9/10 | 9/10 |
| Hexachlorodibenzo-*p*-dioxin (c) [c], μg/kg | | | | |
| 0 | 1/10 | 0/10 | 0/10 | 0/10 |
| 0.1 | 0/10 | 1/10 | 0/10 | 0/10 |
| 1.0 | 0/10 | 0/10 | 0/10 | 0/10 |
| 10.0 | 1/10 | 9/10 | 3/10 | 1/10 |
| 100.0 | 10/10 | 6/10 | 3/10 | 8/10 |
| Octachlorodibenzo-*p*-dioxin (d) [d], % of diet | | | | |
| 0 | 0/12 | 0/12 | 0/12 | 0/12 |
| 0.1 | 0/12 | 0/12 | 0/12 | 0/12 |
| 0.5 | 0/12 | 0/12 | 0/12 | 0/12 |

[a] Sample identified in Table I.
[b] Number affected/total number in group.

similar lesions consisting of: atrophy of germinal centers of the spleen, a paucity of lymphocytes in the bursa of Fabricius, pulmonary edema, interstitial edema of the myocardium, fatty degeneration, and coagulation necrosis of the liver. Many birds died as a result of pulmonary edema.

**Pathology.** Gross necropsy and histological examinations were conducted on relatively few mammals treated with the chlorinated dibenzo-*p*-dioxins. Therefore, the results reported here are incomplete and preliminary. The liver of animals treated with 2,3,7,8-TCDD and HCDD was most consistently affected. Microscopic examination of this organ revealed a highly variable pattern and degree of hepatic necrosis with various degrees of degeneration and regeneration of the hepatocytes, depending upon the post-treatment interval. Necrosis was observed both in the centrilobular and periportal areas. The degree of necrosis of the liver was not sufficient to conclude that it was responsible for death. Hepatic lesions were observed in rats, mice, rabbits, and dogs. In addition to hepatic involvement, other changes observed sporadically include fat necrosis, periarteritis, serous atrophy of fat, and ascites.

## *Discussion and Summary*

The studies reported here confirmed the high toxicity of 2,3,7,8-TCDD. In addition, some perspective of the relative toxicities of

**Gross Lesions Observed in Chicks Treated with Chlorodioxins**

| *Pulmonary Edema* | *Atrophy of Spleen and/or Bursa* | *Liver Swollen and/or Mottled* | *Gizzard Erosions* |
|---|---|---|---|
| 0/10 | 0/10 | 0/10 | 0/10 |
| 0/10 | 0/10 | 0/10 | 0/10 |
| 0/10 | 0/10 | 0/10 | 0/10 |
| 5/10 | 2/10 | 7/10 | 1/10 |
| 5/10 | 0/10 | 6/10 | 0/10 |
| | | | |
| 0/10 | 0/10 | 0/10 | 0/10 |
| 0/10 | 0/10 | 0/10 | 0/10 |
| 0/10 | 0/10 | 0/10 | 0/10 |
| 1/10 | 1/10 | 3/10 | 0/10 |
| 8/10 | 0/10 | 4/10 | 0/10 |
| | | | |
| 0/12 | 0/12 | 0/12 | 2/12 |
| 0/12 | 0/12 | 0/12 | 7/12 |
| 0/12 | 0/12 | 0/12 | 7/12 |

[c] Administered orally as a corn oil: acetone solution.
[d] Fed in the diet (0.1% ≅ 100 mg/kg, 0.5% ≅ 500 mg/kg).

2,7-DCDD, HCDD, and OCDD has been obtained. 2,7-DCDD and OCDD failed to cause death in female rats given oral doses of 1 gram/kg; even larger doses were given to mice without causing death. Limited data suggest that oral doses of approximately 100 mg/kg of HCDD are needed to cause death in male rats. In the teratology study, no deaths occurred following administration of 100 $\mu$g/kg of HCDD to female rats for 10 consecutive days.

2,3,7,8-TCDD is much more toxic than the other chlorodibenzodioxins studied; the $LD_{50}$ ranged from 0.6 $\mu$g/kg in male guinea pigs to 115 $\mu$g/kg in rabbits. Dogs appear to be less sensitive than rabbits. Others have reported 100% mortality in rabbits treated with 10 $\mu$g/kg (*10*) and chick embryos treated with 0.05 $\mu$g/egg (*4*).

Death following treatment with a lethal dose of 2,3,7,8-TCDD is often delayed for several weeks. Among the animals which died following treatment, approximately half the deaths occurred between 13 and 18 days after treatment, with one animal dying as late as 43 days after a single oral dose. In mice and rabbits, there is a marked individual difference in susceptibility to this compound which makes it difficult to conduct acute lethality studies.

If the results of the rabbit eye irritation test can be extrapolated to man, accidental contact of these chlorodibenzodioxins with the eyes should not present a serious threat to vision. However, repeated contact with the skin of small amounts of either 2,3,7,8-TCDD or HCDD may be expected to produce chloracne. Sensitivity to 2,3,7,8-TCDD was recognized by industry years ago, and precautions have been taken to minimize its occurrence and prevent contamination of workers' skin. HCDD is apparently a less potent acnegen than 2,3,7,8-TCDD.

As previously reported, 2,3,7,8-TCDD is highly embryotoxic (*12*). The no-effect level for embryotoxicity was 0.03 $\mu$g/kg/day of 2,3,7,8-TCDD. In contrast to the high embryotoxicity of the symmetrical 2,3,7,8-TCDD, 1,2,3,4-TCDD was not embryotoxic at doses as high as 800 $\mu$g/kg/day (*7*).

Using previously described definitions of teratogenicity and embryotoxicity (*11*), HCDD is teratogenic in the rat at a 100 $\mu$g/kg/day dose level, given orally on days 6 through 15 of gestation. Treatment of pregnant rats with HCDD caused embryotoxicity evidenced by a dose-related decrease in fetal body weight and crown-rump length and an increase in the incidence of fetal resorptions (Table III). Likewise, the incidence of certain soft tissue and skeletal anomalies increased in a manner related to the dose level of HCDD (Table IV). A 0.1 $\mu$g/kg/day dosage of HCDD had no effect on embryonal or fetal development.

OCDD caused embryotoxicity but was not teratogenic at 500 mg/kg/day. OCDD and 2,7-DCDD caused neither teratogenicity nor embryotoxicity at 100 mg/kg/day. Khera and Ruddick (*7*) reported that the administration of 2 mg 2,7-DCDD/kg/day was associated with microscopic myocardial and pericardial lesions in rat fetuses. However, examination of sections of myocardium and pericardium from fetuses of dams treated with 100 mg doses in this study revealed no morphological differences from controls.

Both 2,3,7,8-TCDD and HCDD give positive results in check edema bioassays (Table V). This HCDD result is consistent with a previous report that the HCDD isolated from pentachlorophenol produced chick edema (*4*). These same authors reported that 2,3,7,8-TCDD was extremely toxic in the chick embryo assay but did not report that it produced chick edema.

Pathological changes observed in animals treated with chlorodibenzodioxins were inconsistent from animal to animal and species to species. Hepatic lesions were observed consistently, but the nature, degree, and distribution of the lesions were variable. Changes in organs other than the liver were sporadic and unpredictable. Gross and microscopic examination of tissues after chlorodibenzodioxin treatment did not reveal the cause of death. An in-depth evaluation of the toxicity associated with chronic exposure to the chlorodibenzodioxins is needed.

Isomers of a chlorodibenzodioxin can produce different degrees of toxicity; 2,3,7,8-TCDD is highly embryotoxic and a potent acnegen, but 1,2,3,4-TCDD is neither embryotoxic nor acnegenic.

The toxicity of chlorodibenzodioxins other than those evaluated in this study has not been reported. Purified samples of trichloro-, pentachloro-, and heptachlorodibenzo-*p*-dioxin which are free of tetrachloro- and hexachlorodibenzo-*p*-dioxin need to be synthesized for study. However, heptachlorodibenzo-*p*-dioxin cannot be highly toxic since studies on octachlorodibenzo-*p*-dioxin containing several percent of heptachlorodibenzo-*p*-dioxin have tested the same as the pure product.

Studies on the chlorodibenzodioxins have led to the following conclusions: (1) 2,7-dichlorodibenzo-*p*-dioxin and octachlorodibenzo-*p*-dioxin have a low acute toxicity; (2) 2,3,7,8-tetrachlorodibenzo-*p*-dioxin has an unusually high toxicity; (3) hexachlorodibenzo-*p*-dioxin is highly toxic but less toxic than 2,3,7,8-tetrachlorodibenzo-*p*-dioxin; (4) all chlorodibenzodioxins are not alike in their toxicological properties. Isomers of the same dibenzo-*p*-dioxin vary in toxicological properties, making it important to identify them specifically.

## *Acknowledgment*

The authors are grateful to J. E. Bourne, P. A. Keeler, and R. W. Lisowe for their assistance in all aspects of this study.

## *Literature Cited*

1. *Chem. Eng. News* (1967) **45,** 10.
2. Dawson, A. B., *Stain Technol.* (1926) **1,** 123–124.
3. Herxheimer, K., *Muenchen. Med. Wochenschr.* (1899) **46,** 278.
4. Higginbotham, G. R., Huang, A., Firestone, D., Verrett, J., Ress, J., Campbell, A. D., *Nature* (1968) **220,** 702.
5. Hofman, H. T., *Arch. Exp. Pathol. Pharmakol.* (1957) **232,** 228–230.
6. Jones, E. L., Krizek, H., *J. Invest. Dermatol.* (1962) **39,** 511–517.
7. Khera, K. S., Ruddick, J. A., ADVAN. CHEM. SER. (1973) **120,** 70.
8. Kimmig, J., Schulz, K. H., *Dermatologica* (1957) **115,** 540–546.
9. Litchfield, J. T., Wilcoxon, F., *J. Pharmacol. Exp. Ther.* (1949) **96,** 99–113.
10. Milnes, M. H., *Nature* (1971) **232,** 395–396.
11. Schwetz, B. A., Sparschu, G. L., Gehring, P. J., *Food Cosmet. Toxicol.* (1971) **9,** 801–817.
12. Sparschu, G. L., Dunn, F. L., Rowe, V. K., *Food Cosmet. Toxicol.* (1971) **9,** 405–412.
13. Steel, R. G. D., Torrie, H. H., "Principles and Procedures of Statistics," pp. 73–75, 81, 347, 349, 366–372, McGraw-Hill, New York, 1960.
14. Thompson, W. R., *Bacteriol. Rev.* (1947) **11,** 115–145.
15. Weil, C. S., *Biometrics* (1952) **8,** 249–263.
16. Wilson, J. G., "Teratology Principles and Techniques," J. G. Wilson, T. Warkany, Eds., pp. 262–277, University of Chicago Press, Chicago, 1965.

RECEIVED October 24, 1972.

# 8

# Polychlorodibenzo-*p*-dioxins: Perinatal Effects and the Dominant Lethal Test in Wistar Rats

K. S. KHERA and J. A. RUDDICK

Health Protection Branch, Department of National Health and Welfare, Tunney's Pasture, Ottawa, Ontario K1A 0L2, Canada

*Reduced embryonic viability and fetal weight, were found in offspring from Wistar rats treated orally during days 6–15 of gestation with 1–16 μg/kg/day of 2,3,7,8-tetrachlorodibenzo-*p*-dioxin (8-D). Viable fetuses from 0.25 μg/kg/day and higher dose groups had a dose-related incidence of visceral anomalies characterized by edema and hemorrhage. At the 0.125 μg/kg/day level, no fetopathic effect was apparent. Postnatally, the servival, body weight gain, and reproductive ability of the progeny were adversely affected following maternal dosing with 1–0.5 μg/kg/day of 8-D. Specific activity occurred in the fetus, placenta, liver, kidney, brain, and fat of the gravida. Additional dosing with 4-D, 2-D, 2,3-dichlorodibenzo-*p*-dioxin, or 2-chlorodibenzo-*p*-dioxin during gestation produced no significant effects.*

A consistent pericardial edema in chickens gave rise to the term chick edema disease (chick edema factor) (*1*). Two known outbreaks of the disease in the broiler industry resulted in a great loss of chickens. A lipid residue from the manufacturing fatty acids, being used as a feed ingredient, was a principal source of the toxic substance. Contamination of the lipid component with polychlorodibenzo-*p*-dioxins was attributed as the causal agent.

Industrial workers involved in chlorinated aromatic production including chlorophenol suffered dioxin-induced chloracne (*2, 3*). Chloracne and other serious health disturbances have been attributed to polychlorodibenzo-*p*-dioxins in workers involved in manufacturing 2,4,5-T (*4, 5*). Dioxins are toxic to chick embryos, guinea pigs, rabbits, and monkeys (*6, 7, 8, 9, 10*).

Polychlorodibenzo-*p*-dioxins contain from one to eight chlorine atoms with a theoretical possibility of forming 75 compounds, many of them being isomers. Thus far, ten of these dioxins have been identified as contaminants in food, vegetable oils, animal feeds, technical grades of chlorophenols, and the herbicide, 2,4,5-T (*11, 12*). Such findings warranted a teratologic and mutagenic evaluation of identified members of the dioxin family.

## *Experimental*

Female Wistar rats treated orally with 0.125–16 μg/kg/day of 2,3,7,8-tetrachlorodibenzo-*p*-dioxin (8-D), 50–800 μg/kg/day of 1,2,3,4-tetrachlorodibenzo-*p*-dioxin (4-D), and 250–2000 μg/kg/day of 2,7-dichlorodibenzo-*p*-dioxin (2-D) and 1,000–2,000 μg/kg/day of 2,3-dichlorodibenzo-*p*-dioxin or 2-chlorodibenzo-*p*-dioxin during 6–15 days of gestation (first morning of a positive vaginal smear was designated as day 1) were assessed for prenatal effects of the compounds on their progeny.

The test compound was dissolved in a minimal volume of anisole, which was in turn diluted in corn oil. The anisole concentration never exceeded 1.5% (v/v). Treated controls were dosed with an equivalent anisole–corn oil concentration. Untreated controls (not subjected to experimental manipulation) were also included in the experiments. Females were weighed on days 1, 6–15, and 22 of gestation; maternal postmortem weights were also recorded. The females in the prenatal studies were killed, and their uteri examined on the 22nd day of gestation. The number of viable and dead fetuses were noted. Viable fetuses were weighed and processed for skeletal, visceral, or microscopic examinations. Visceral examination of fetuses was conducted on gross sections (3–4 mm thick) cut on the saggital plane. For microscopy mid-saggital sections from whole fetuses were stained with haematoxylin–phloxine–saffron.

The postnatal studies were conducted only on 8-D, 4-D, and 2-D. Dams, given oral doses of these dioxins, were allowed to litter. Values for stillborn were less certain as some of the stillborn might have been missed, owing to cannibalism. The pups were examined frequently for viability, body weight gain, gross defects, and signs of abnormal behavior. The progenies on reaching maturity were mated within treatment groups. Values for female and male fertility (ability to become pregnant or induce pregnancy) and numbers of viable fetuses, dead fetuses, resorption sites, and corpora lutea were analyzed for possible effects.

Uniformly labeled $^{14}C$-8-D with a specific activity of 2.99 μc/mg was administered orally to pregnant females at 2 μg/kg/day from 6–15 days of gestation. Three females were sacrificed on alternate days during days 6–20 of pregnancy. Triplicate samples of fetus, placenta, blood, brain, abdominal fat, and sartorius muscle were procured from each female. The samples were dissolved in 1 ml of Soluene (Packard Instruments) to which 15 ml of Aquasol were added. Each sample vial was counted for 30 min in a Nuclear Chicago Mark I liquid scintillation counter.

A second experiment with the labeled 8-D (same activity as in the above experiment) was designed to determine whether the label crossed

**Table I. 2,3,7,8-Tetrachlorodibenzo-**

| *Experiment No.* | *Dose-Group (μg/kg)* | *Maternal Toxicity* | *No. of Dams* | *Alive Fetuses Ave/1* |
|---|---|---|---|---|
| I | Untreated control | − | 13 | 11.5 |
| | Treated control | − | 12 | 9.8 |
| | 16 | + | 7 | 0 |
| | 8 | + | 9 | 0 |
| | 4 | + | 7 | 0 |
| | 2 | + | 13 | 6.0 |
| | 1 | + | 15 | 6.5 |
| II | Untreated control | − | 13 | 10.7 |
| | Treated control | − | 13 | 11.0 |
| | 1 | ± | 11 | 9.3 |
| | 0.5 | − | 13 | 10.5 |
| | 0.25 | − | 13 | 10.9 |
| | 0.125 | − | 15 | 10.6 |

[a] Number of abnormal fetuses/total number examined.

the placenta and entered the fetus. The compound was given as a single oral dose of 200 μg/kg to each of the two dams on gestation day 16, 17, or 18. The dams were killed 6 hours after dosing, and triplicate samples were obtained except the muscle.

The fetuses and placentae were homogenized with a hand grinder in 3 ml of methanol:chloroform (2:1), and the homogenate was transferred to a stoppered centrifuge tube. The homogenizer was rinsed twice with 1 ml of the methanol:chloroform mixture each time, and the rinses were combined with the homogenate. The tubes were set in a water bath at 50°C for 1 hour. Then, the tubes were centrifuged at 2000 *g* for 10 min. The supernatant and precipitate were transferred to separate scintillation counting vials. Once the methanol and chloroform were evaporated by heating in an oven at 40°C, 1 ml of Soluene followed by 15 ml of Aquasol was added to each vial and counted as described previously. The maternal tissues were handled as in the preceding experiment.

A dominant lethal test was conducted with groups of male rats dosed orally with 4, 8, or 12 μg/kg/day of 8-D for seven consecutive days after which seven sequential mating trials were conducted in the surviving males. At a mating trial each male was caged with two untreated virgin females for 5 days. The females were killed 9 days after separation from the males, and viable embryos, resorption sites, and corpora lutea were counted. At the end of the mating schedule the surviving males were killed, and their testes and epididymis were examined histologically.

### *Experimental Results*

2,3,7,8-TETRACHLORODIBENZO-*p*-DIOXIN (8-D). PRENATAL EFFECTS. Data from 2 experiments conducted with 0.125–16 μg/kg/day doses of

**p-Dioxin: Prenatal Data**

| *Dead and Resorbed Fetuses Ave/1* | *Fetal Weight (g) Ave* | *Skeleton Anomalies*[a] *(%)* | *Visceral Lesions*[a] | |
|---|---|---|---|---|
| | | | *Gross* | *Microscopic* |
| 0.2 | 4.68 | 8/116 (7) | 0/16 | 0/10 |
| 0.8 | 4.77 | 9/89 (10) | 0/18 | 0/10 |
| 9.3 | — | — | | |
| 11.8 | — | — | | |
| 11.0 | — | — | | |
| 6.7 | 3.31 | 7/57 (12) | 11/14 | 3/7 |
| 3.7 | 4.17 | 7/80 (7) | 7/12 | 3/6 |
| 0.6 | 4.82 | 5/107 (5) | ND[b] | 0/13 |
| 0.9 | 4.51 | 21/116 (18) | 0/16 | 0/11 |
| 0.8 | 4.10 | 6/81 (7) | 2/11 | 3/10 |
| 1.1 | 4.46 | 10/105 (10) | ND | 3/31 |
| 0.2 | 4.79 | 6/109 (6) | ND | 1/33 |
| 0.1 | 4.64 | 3/121 (2) | ND | 0/38 |

[b] Not done.

8-D are summarized in Table I. Doses of 4 μg/kg/day or more produced maternal toxicity and 100% embryonic lethality. Two and 1 μg/kg/day treatments were also toxic for the mothers and resulted in reduced number of viable fetuses per litter and reduced fetal weights. Lower doses

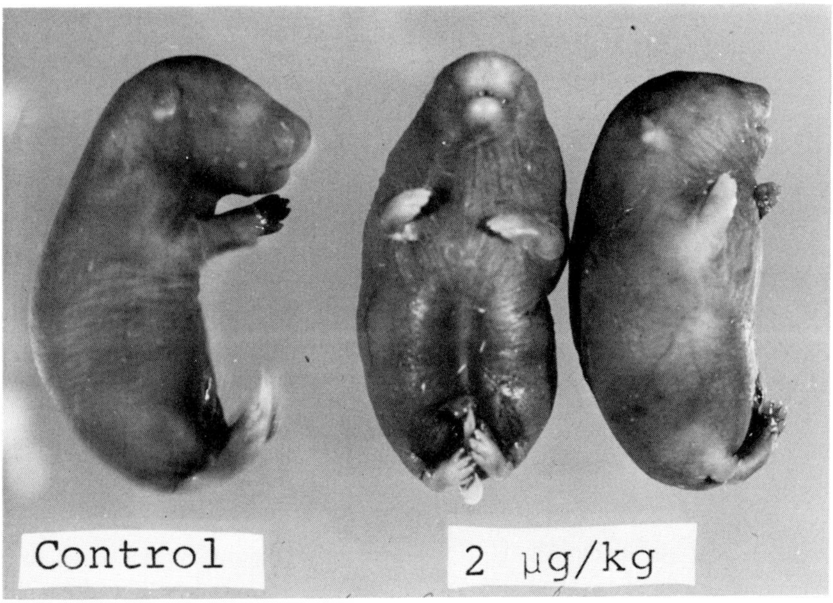


*Figure 1. Subcutaneous fetal edema following maternal dosing with 2 mg/kg/day of 2,3,7,8-tetrachlorodibenzo-p-dioxin*

(0.5–0.125 μg/kg/day), however, had no adverse effect on these parameters. The incidence of skeletal anomalies, presented as the number of malformed fetuses *vs.* the total number of fetuses examined, is given in Table I. Wavy ribs, lumbar ribs, and various sternal defects already reported for this strain of rats (*13*) were observed in the test as well as control groups. At the dose range of 2.0–0.125 μg/kg/day the incidence was comparable with the control. Visceral lesions in test fetuses consisted of subcutaneous edema of head, neck, and trunk (Figure 1) and hemorrhages in the intestinal lumen (Figure 3). The hemorrhages were comparable with previous observations (*14, 15*), but further hemorrhages in the brain cavities and subcutaneous tissues were also noted. These lesions were observed in gross and histologic sections of fetuses from the 0.25–2 μg/kg groups; their incidence was dose-related. No fetopathic effect was noticed at the 0.125 μg/kg/day level. None of the lesions were observed in the controls.

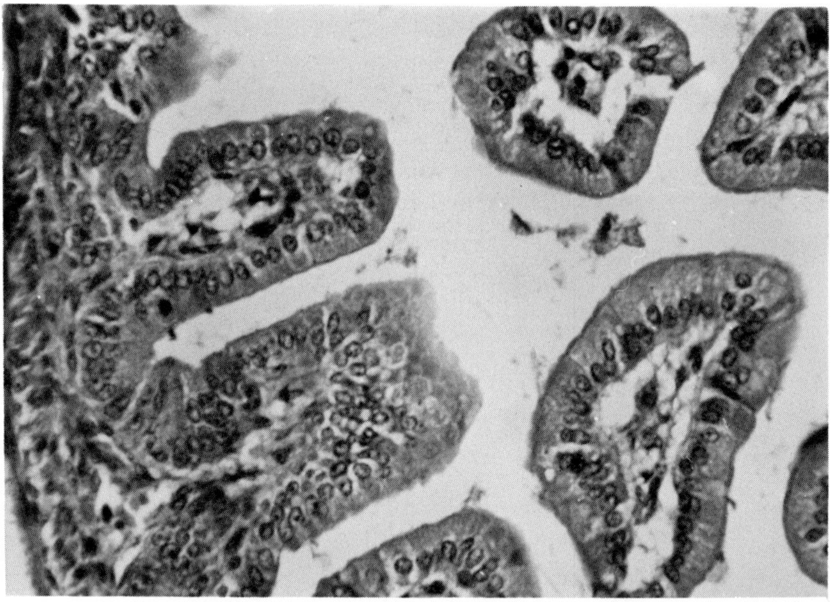

*Figure 2. Control, section of rat intestine (100×)*

Postnatal Effects. Progeny survival and average pup weight at birth and at weaning are tabulated in Table II. The average number of viable pups littered and pup weight at birth indicated a dose related reduction at the 0.25, 0.5, and 1 μg/kg doses. Progeny viability until weaning (day 21) and average pup weight of the weanlings from 0.5 and 1 μg/kg groups were reduced as compared with the controls. No pups delivered from the 1 μg/kg treated mothers survived until weaning

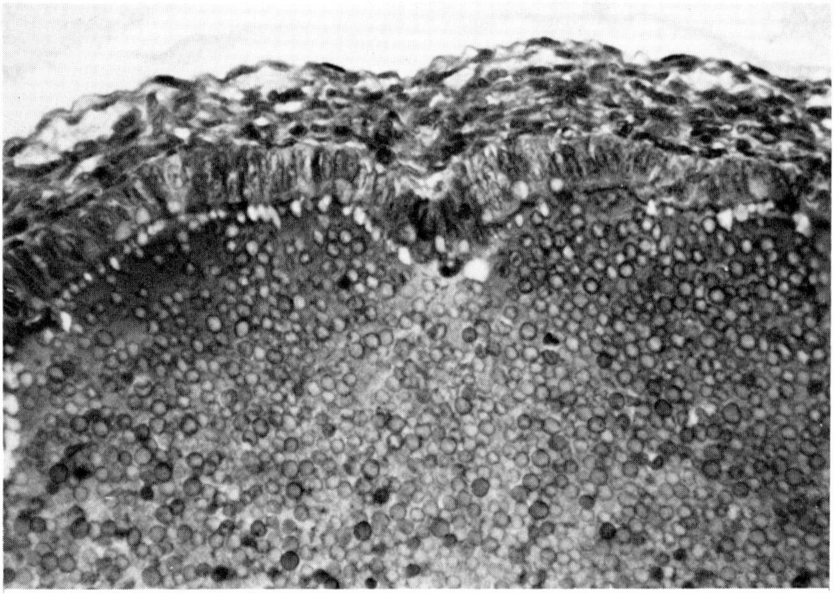

*Figure 3. Section of intestine from fetus whose mother was dosed orally with 1 μg/kg/day of 2,3,7,8-tetrachlorodibenzo-p-dioxin during gestation days 6-15 (Note hemorrhage in intestinal lumen) (100×)*

while 60% of the pups from the 0.5 μg/kg dosed mothers lived to day 21. Treatment with 0.125 μg/kg had no adverse postnatal effect on the progeny (Table II).

Postnatal mortality in littered pups was investigated further by transferring 42 newly born pups from five mothers orally treated with 1 μg/kg during 6–15 days of gestation to five control dams. Forty-six neonates from the control mothers were transferred to the five treated 8-D dams. The pups were reared by the exchanged mothers until weaning. Thirty-six of the 42 pups from the 8-D group died, whereas only two of the 46 control pups died. The postnatal lethality, therefore, seemed to be a late expression of *in utero* embryonic damage induced by 8-D and not resulting from excretion of toxic substances in the maternal milk.

The progeny, mated within the experimental group, had a decreased number of pregnancies and reduced litter size at the 0.5 μg/kg while no effect was observed in 0.25 μg/kg progeny.

LOCALIZATION OF $^{14}$C-2,3,7,8-TETRACHLORODIBENZO-*p*-DIOXIN IN TISSUES OF PREGNANT RATS. The radioactivity, expressed as micrograms of 8-D per gram wet weight was detected in liver, fat, and brain but not in blood, kidney, sartorius muscle, embryo, and placenta. On day 8 of gestation 0.07 μg/gram of 8-D was observed in the liver (Figure 4). The amount increased to 0.21 μg/gram on the 12th day and remained constant

**Table II. 2,3,7,8-Tetrachlorodibenzo-**

| *Dose-Group (μg/kg)* | *Maternal Toxicity* | *No. of Dams* | *No. of Viable Pups Littered (Ave./1)* | *No. Stillborn (Ave./1)* | *Pup Weight at Birth Ave. (g)* |
|---|---|---|---|---|---|
| Untreated control | − | 14 | 12.0 | 0.4 | 6.38 |
| Treated control | − | 14 | 11.6 | 0.2 | 6.54 |
| 1 | + | 14 | 5.9 | 2.4 | 5.18 |
| 0.5 | + | 14 | 9.9 | 1.0 | 5.70 |
| 0.25 | − | 14 | 10.2 | 0.4 | 5.97 |
| 0.125 | − | 15 | 10.9 | 0.3 | 6.51 |

until the 20th day of gestation (0.18 μg/gram). The amount in the fat reached 0.028 μg/gram on the 12th day and increased to 0.052 μg/gram on the 20th day. The brain accumulated label until the 10th day of gestation (0.029 μg/gram), dropped on the 12th day to 0.016 and then increased to 0.028 μg/gram on the 20th day. No significantly increased

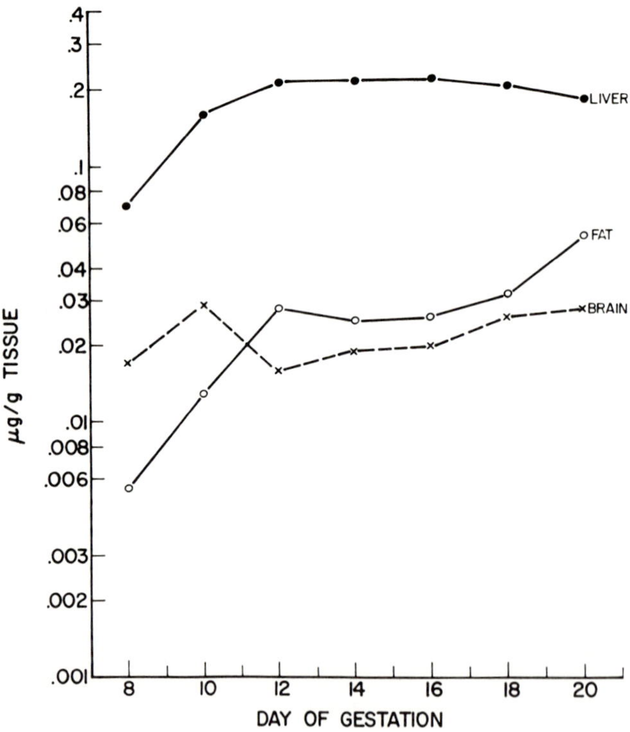


*Figure 4. Distribution of* [14]*C-2,3,7,8-tetrachlorodibenzo-p-dioxin in pregnant Wistar rats treated orally from 6-15 days of gestation with 2 μg/kg/day*

**p-Dioxin: Postnatal Effects on Progeny**

| *No. of Pups Allowed to Survive* | *Mortality During Ages (Days)* | | | | *Survival on Day 21 (%)* | *Pup Weight on Day 21 Ave. (g)* |
|---|---|---|---|---|---|---|
| | 0–1 | 1–2 | 3–10 | 11–21 | | |
| 128 | 0 | 0 | 1 | 4 | 123 (96) | 31.99 |
| 149 | 0 | 0 | 5 | 13 | 131 (88) | 30.22 |
| 40 | 20 | 2 | 13 | 5 | 0 (0) | — |
| 139 | 14 | 6 | 11 | 25 | 83 (60) | 24.99 |
| 143 | 5 | 0 | 2 | 7 | 129 (90) | 30.96 |
| 164 | 1 | 0 | 4 | 6 | 153 (93) | 29.91 |

radioactivity in the tissues was detected after 12 days of gestation (Figure 4) despite daily dosing being continued until the 15th day of gestation. Further data on rates of absorption, metabolism, and excretion, using 8-D of higher specific activity will be needed to explain this observation.

Table III presents the amounts of $^{14}$C-8-D (μg/gram) from the second label experiment. On gestation day 16, 17, and 18, the average values of 0.011, 0.015, and 0.016 μg/gram were observed for the fetus while 0.025, 0.038, and 0.041 μg/gram were obtained for the placenta. The average fetal weights were 277, 464, and 801 mg and the placental weights were 218, 300, and 341 mg on days 16, 17, and 18, respectively. The dams (300 grams) received approximately 60 μg/kg of 8-D as a single dose (at the rate of 200 μg/kg) on these gestation days. With these data it was estimated that the percentage of the dose present was 0.005, 0.012, and 0.021 for the fetus and 0.009, 0.019, and 0.023 for the placenta on days 16, 17, and 18, respectively.

DOMINANT LETHAL ASSAY. All doses of 8-D used in the dominant lethal study were toxic for the male rats. The highest dose, 12 μg/kg/day, was lethal for 20/20 males (mean survival, 17.7 days), 8 μg/kg/day killed 11/20 males (mean survival, 20.1 days), and 4 μg/kg/day was lethal for 2/20 (mean survival, 36.5 days). There were no mortalities in the control group.

Reproductive values indicated no dominant lethal mutations during 35 days post-treatment (Table IV). The period examined corresponded to postmeiotic stages of spermatogenesis. The incidence of pregnancies at all mating trials in the treated groups remained lower than the control.

Testes from the males examined 43 days after the 8-D treatment appeared normal. However, the epididymis was involved in an inflammatory process with sperm granulomas formation. The granulomatous epididymal lesion resembled the lesion seen in auto-immune reaction following bacterial infections or tissue response to foreign bodies (*16, 17,*

**Table III. Tissue Distribution in Dams Following a Single Oral Dose of 200 μg/kg of $C^{14}$-2,3,7,8-Tetrachlorodibenzo-*p*-Dioxin on Gestation Day 16, 17, or 18 (μg/gram of tissue)**

| | *Gestation Day* | | |
|---|---|---|---|
| | *16* | *17* | *18* |
| Liver | 0.339 ± 0.015 | 0.339 ± 0.019 | 0.275 ± 0.020 |
| Kidney | 0.035 ± 0.008 | 0.058 ± 0.006 | 0.049 ± 0.005 |
| Fat | 0.105 ± 0.025 | 0.113 ± 0.010 | 0.089 ± 0.008 |
| Brain [a] | 0.040 ± 0.008 | 0.047 ± 0.003 | 0.041 ± 0.002 |
| Blood | 0.025 ± 0.011 | 0.019 ± 0.009 | 0.010 ± 0.003 |
| Fetus | 0.011 ± 0.003 | 0.015 ± 0.001 | 0.016 ± 0.001 |
| Placenta | 0.025 ± 0.006 | 0.038 ± 0.004 | 0.041 ± 0.003 |
| Fetus ppte [b] | — [c] | — | — |
| Placenta ppte | — | — | — |

[a] Standard error in blood is 50% while all other samples are from 3–10%.
[b] Precipitate.
[c] No label observed.

*18*). In addition, the ductal mucosa in the caudal region was hyperplastic, being lined by 4–5 rows of cells. Epidermal hyperplasia following topical application of 8-D on rabbit's ear is known (*19*).

**1,2,3,4-Tetrachlorodibenzo-*p*-Dioxin.** 1,2,3,4-Tetrachlorodibenzo-*p*-dioxin tested for prenatal effects using 50–800 μg/kg/day during 6–15 days of gestation showed no maternal toxicity (Table V). Average values for viable fetuses, dead fetuses, and fetal weight were within the control range. Microscopic examination did not reveal effects at the two high dose levels of 400 and 800 μg/kg/day. Incidence of gross visceral and skeletal anomalies were within control limits (Table V).

Postnatally, until weaning the viability and mean body weight of progeny from dams treated with 800 μg/kg of 1,2,3,4-tetrachlorodibenzo-

**Table IV. Pregnancies and Viable Embryos and 2,3,7,8-Tetrachlorodibenzo-**

| *Mating Trial, No.* | *Mating Intervals After Dosing (days)* | *Viable Embryos Doses (μg/kg)* | | |
|---|---|---|---|---|
| | | *0* | *4* | *8* |
| 1 | 0–5 | 12.1 | 12.8 | 8.7 |
| 2 | 5–10 | 9.1 | 10.7 | 14 |
| 3 | 10–15 | 10.0 | 11.5 | 6 [b] |
| 4 | 15–20 | 10.2 | 10.3 | 12.3 |
| 5 | 20–25 | 12.4 | 12.4 | 12.5 |
| 6 | 25–30 | 10.7 | 12.0 | 13.0 |
| 7 | 30–35 | 13.2 | 11.2 | 11.3 |

[a] Values based on 40 females mated to 20 males (in each of the 0 and 4 μg/kg groups) and 12 females mated to 6 of the surviving males (8 μg/kg).

*p*-dioxin were within control range (Table VI). Likewise, no effect was noticed on male or female fertility and embryonic viability after mating the progeny which became adult.

**2,7-Dichlorodibenzo-*p*-Dioxin.** Prenatal data from rats treated during gestation with 250–2000 μg/kg/day of 2,7-dichlorodibenzo-*p*-dioxin are given in Table VI. The average litter size distributions of dead fetuses per litter and the average fetal weight in the test groups were not significantly different from the control. The incidence of skeletal malformation *vs.* the total number examined for the various dose groups indicated no significant deviation from control values (Table VI). The types of malformations were the same as described above.

Microscopic examination of the heart revealed edematous separation of myofibrils that had resulted in compression thinning and fragmentation of myofibres. Myofibre outlines were less distinct, and there was loss of acidophilic staining. Mitotic figures were rare, indicating that growth of the cardiac tissue was suppressed. The incidence of cardiac lesions is given (Table VI).

The postnatal effects of maternal treatment with 2,7-dichlorodibenzo-*p*-dioxin are summarized in Table VII. No significant changes were found. The progeny on becoming adult were mated within the treatment groups. No effects on male or female fertility, embryonic viability, and total implantation/corpora lutea ratio were observed.

**2,3-Dichlorodibenzo-*p*-Dioxin.** Maternal treatment with 1,000 and 2,000 μg/kg/day of 2,3-dichlorodibenzo-*p*-dioxin during 6–15 days of pregnancy had no effect on fetal viability, average fetal weight, and normal incidence of skeletal anomalies (Table VIII). None of the gravidas at the two doses manifested any toxicity sign.

**2-Chlorodibenzo-*p*-Dioxin.** Prenatal parameters at dose levels of 1,000 and 2,000 μg/kg/day of 2-chlorodibenzo-*p*-dioxin are given (Table

**Resorption Sites per Pregnancy in Females Mated to *p*-Dioxin Treated Males**[a]

| *Resorption Sites Doses (μg/kg)* | | | *Percent Pregnancy Doses (μg/kg)* | | |
|---|---|---|---|---|---|
| *0* | *4* | *8* | *0* | *4* | *8* |
| 0 | 0 | 0 | 30 | 15 | 8 |
| 2.1 | 0 | 0 | 50 | 15 | 3 |
| 0 | 0 | 0 | 43 | 10 | 2 |
| 0 | 0 | 0 | 43 | 18 | 25 |
| 0.1 | 0.1 | 0 | 80 | 28 | 25 |
| 0 | 0 | 0 | 73 | 43 | 34 |
| 0.1 | 0.1 | 0 | 83 | 55 | 25 |

[b] Embryos only from one pregnancy.

**Table V. 1,2,3,4-Tetrachlorodibenzo-*p*-**

| *Dose-Group (μg/kg)* | *Maternal Toxicity* | *No. of Dams* | *Alive Fetuses Ave/1* |
|---|---|---|---|
| Control (untreated) | — | 10 | 12.0 |
| Control (treated) | — | 13 | 12.5 |
| 800 | — | 15 | 11.7 |
| 400 | — | 14 | 12.6 |
| 200 | — | 15 | 12.1 |
| 100 | — | 14 | 11.4 |
| 50 | — | 15 | 12.2 |

[a] Number of anomalous fetuses/total number examined.

**Table VI. 2,7-Dichlorodibenzo-**

| *Dose-Group (μg/kg)* | *Maternal Toxicity* | *No. of Dams* | *Alive Fetuses Ave./L* | *Dead and Resorbed Fetuses Ave./L* | *Fetal Weight Ave. (g)* |
|---|---|---|---|---|---|
| Untreated control | — | 14 | 10.9 | 0.9 | 4.48 |
| Treated control | — | 15 | 11.0 | 0.3 | 4.70 |
| 2,000 | — | 13 | 10.5 | 0.6 | 4.85 |
| 1,000 | — | 15 | 11.2 | 0.7 | 4.43 |
| 500 | — | 15 | 11.1 | 0.9 | 4.35 |
| 250 | — | 13 | 11.0 | 0.3 | 4.77 |

[a] Number of abnormal fetuses/total number examined.

**Table VII. 1,2,3,4-Tetrachlorodibenzo-*p*-Dioxin and**

| *Chlorodibenzo-p-dioxin* | *Dose-Group (μg/kg)* | *Maternal Toxicity* | *No. of Dams* |
|---|---|---|---|
| | Treated control | — | 11 |
| 1,2,3,4-Tetra-. | 800 | — | 12 |
| 2,7-Dichloro-. | 2,000 | — | 12 |
| | 1,000 | — | 13 |
| | 500 | — | 13 |
| | 250 | — | 15 |

**Dioxin: Prenatal Effects on Rat Fetus**

| *Dead and Resorbed Fetuses Ave/1* | *Fetal Weight (g), Ave* | *Skeletal Anomalies*[a] | *Visceral Lesions*[a] | |
|---|---|---|---|---|
| | | | *Microscopic* | *Gross* |
| 0.4 | 4.65 | 5/81 | ND[b] | 1/31 |
| 0.2 | 4.71 | 7/101 | 0/12 | 2/41 |
| 0.6 | 4.89 | 7/107 | 0/13 | 2/53 |
| 0.5 | 4.48 | 13/118 | 0/14 | 1/41 |
| 0.3 | 4.57 | 15/108 | ND | 1/45 |
| 0.9 | 4.66 | 16/107 | ND | 1/51 |
| 0.4 | 4.63 | 8/115 | ND | 1/46 |

[b] Not done.

***p*-Dioxin: Prenatal Effects**

| *Skeleton Anomalies*[a] *(%)* | *Microscopic Examination* | | |
|---|---|---|---|
| | *No. of Fetuses Examined* | *No. of Fetuses with Lesions of* | |
| | | *Myocardium* | *Hydro pericardium* |
| 13/102 (13) | ND | | |
| 13/109 (12) | 22 | 0 | 3 |
| 6/91 (7) | 18 | 7 | 2 |
| 8/112 (7) | 24 | 2 | 5 |
| 20/101 (18) | 19 | 0 | 8 |
| 8/103 (8) | 22 | 0 | 5 |

[b] Not done.

**2,7-Dichlorodibenzo-*p*-Dioxin: Postnatal Effects on Progeny**

| *No. of Viable Pups Littered Ave./L* | *Stillborn + Dead after Birth Ave./L* | *Pup Weight at Birth Ave. (g)* | *Survival on Day 21 (%)* | *Pup Weight on Day 21 Ave. (g)* |
|---|---|---|---|---|
| 12.5 | 0.1 | 6.52 | 96 | 32.52 |
| 11.7 | 0.4 | 6.36 | 95 | 33.22 |
| 10.3 | 0 | 6.69 | 93 | 33.22 |
| 9.6 | 0.2 | 6.61 | 99 | 34.24 |
| 11.1 | 0.2 | 6.65 | 87 | 36.65 |
| 11.7 | 0.1 | 6.71 | 98 | 33.72 |

**Table VIII. 2,3-Dichlorodibenzo-*p*-Dioxin**

| *Test Compound* | *Dose-Group (μg/kg)* | *Maternal Toxicity* | *No. of Dams* |
|---|---|---|---|
| | Treated control | — | 12 |
| 2,3-Dichlorodibenzo-*p*-dioxin | | | |
| | 2,000 | — | 11 |
| | 1,000 | — | 12 |
| 2-Chlorodibenzo-*p*-dioxin | | | |
| | 2,000 | — | 12 |
| | 1,000 | — | 11 |

[a] Number of malformed fetuses/total number examined.

VIII). Except for slight increase in resorption sites, all fetal measures were within control range.

### *Conclusion*

Oral treatment of pregnant dams with 0.25 μg (or more) /kg/day of 2,3,7,8-tetrachlorodibenzo-*p*-dioxin for 10 days during gestation resulted in adverse effects on rat development. No adverse effects were seen at the 0.125 μg/kg/day. When $^{14}C$-2,3,7,8-tetrachlorodibenzo-*p*-dioxin (2.99 μc/mg) was given at 2 μg/kg/day there was activity, primarily in liver and to a lesser extent, in fat and brain. When a single oral dose of 200 μg/kg was administered on gestation days 16, 17, or 18 and was followed 6 hours later with tissue sampling, the label was also observed in the fetus and placenta. Placenta had approximately twice as much label as the fetus.

In male rats 2,3,7,8-tetrachlorodibenzo-*p*-dioxin induced hyperplastic changes and sperm granulomas in the epididymis, but no apparent lethal mutations were noted during post-meiotic phases of spermatogenesis.

1,2,3,4-Tetrachlorodibenzo-*p*-dioxin elicited no apparent prenatal or postnatal effects when doses of up to 800 μg/kg/day were given orally for 10 days of gestation. Treatment with 250–2000 μg/kg/day of 2,7-dichlorodibenzo-*p*-dioxin (99% purity) had no significant effect on prenatal and postnatal measures of toxicity but caused a low incidence of cardiac lesions. 2,3-Dichlorodibenzo-*p*-dioxin and 2-chlorodibenzo-*p*-dioxin up to 2000 μg/kg/day had no adverse effect on survival, average weight, and skeleton of term fetuses.

**and 2-Chlorodibenzo-*p*-Dioxin: Prenatal Data**

| *Alive Fetuses Ave./L* | *Dead and Resorbed Fetuses Ave./L* | *Fetal Weight (g) Ave.* | *Skeletal Anomalies*[a] (%) |
|---|---|---|---|
| 11.3 | 0.5 | 5.07 | 14/92 (15) |
| | | | |
| 11.4 | 0.8 | 5.11 | 5/84 (6) |
| 10.9 | 0.8 | 5.02 | 5/88 (6) |
| | | | |
| 10.1 | 1.1 | 5.01 | 9/80 (11) |
| 11.5 | 0.8 | 4.86 | 6/83 (7) |


### *Acknowledgments*

We would like to thank A. E. Pohland, Food and Drug Administration, Washington, for supplying us with the dioxins and G. E. Lynn, Dow Chemical Co., Michigan for providing us with the radiolabeled compound. We also thank L. L. Whitta and J. R. Tanner for their technical assistance and H. L. Trenholm and C. J. Paul for their cooperation.


### *Literature Cited*


1. Horwitz, W., "Official Methods of Analysis of the Association of Official Agricultural Chemists," Section 26.087 (1965).
2. Schwartz, L., Tulipan, L., Birmingham, D., "Occupational Diseases of the Skin," p. 336, Lea and Febiger, Philadelphia, 1967.
3. Olivier, N. E., *Arch. Dermatol.* (1969) **99,** 127.
4. Kimmig, J. Schulz, K. H., *Dermatologica* (1957) **115,** 540.
5. Poland, A. P., Smith, D., *Arch. Environ. Health* (1971) **22,** 316.
6. Higginbotham, G. R., Huang, A., Firestone, D., Verrett, J., Ress, J., Campbell, A. D., *Nature* (1968) **220,** 702.
7. Dugois, P., Marechal, J., Colomb, L., *Arch. Mal. Prof.* (1958) **19,** 626.
8. Friedmann, L., Firestone, D., Horwitz, W., Banes, D., Anstead, M., Shue, G., *J. Ass. Offic. Anal. Chem.* (1959) **42,** 129.
9. Allen, J. R., Carstens, L. A., *Lab. Invest.* (1966) **15,** 970.
10. Allen, J. R., Carstens, L. A., *Amer. J. Vet. Res.* (1967) **28,** 1513.
11. Verrett, J., "Effects of 2,4,5-T on Man and the Environment," p. 190, Hearings before the Subcommittee on Energy, Natural Resources, and the Environment of the Committee on Commerce (U.S. Senate), April 7 and 15, 1970, U.S. Government Printing Office, Washington, D.C. (1970).
12. Campbell, A. D., Firestone, D., unpublished.
13. Khera, K. S., McKinley, W. P., *Toxicol. Appl. Pharmacol.* (1972) **22,** 14.
14. Sparschu, G. L., Dunn, F. L., Rowe, V. K., *Fd. Cosmet. Toxicol.* (1971) **9,** 405.
15. Collins, T. F. X., personal communication.

16. Jebsen, J. L., Hartley, W. J., McClure, T. J., McFarlane, D., *New Zeal. Vet. J.* (1955) **3,** 100.
17. Biberstein, E. L., McGowan, B., Olander, H., Kennedy, P. C., *Cornell Vet.* (1964) **54,** 27.
18. Shott, L., Young, S., *Cornell Vet.* (1971) **61,** 281.
19. Jones, E. L., Krizek, H., *J. Invest. Dermatol.* (1962) **39,** 511.

RECEIVED March 1, 1972.

9

# Excretion and Tissue Distribution of 2,3,7,8-Tetrachlorodibenzo-*p*-dioxin in the Rat

WALTER N. PIPER, JAMES Q. ROSE, and P. J. GEHRING

Chemical Biology Research, The Dow Chemical Co., Midland, Mich. 48640

*The tissue distribution and excretion of $^{14}C$ activity have been determined for rats given a single oral dose of tetrachlorodibenzo-*p*-dioxin-$^{14}C$. The half-life for the disappearance of $^{14}C$ activity from the body is 17.4 ± 5.6 days. Examining various tissues 3, 7, and 21 days following administration revealed that the concentration of $^{14}C$ activity in liver and fat was about 10-fold greater than that found in any other tissue. In the liver the percentages of the dose per gram at 3, 7, and 21 days were 3.18, 4.49, and 1.33, respectively. Comparable values for fat were 2.60, 3.22, and 0.43. A total of 53.2% of the dose was eliminated* via *the feces and 13.2 and 3.2%* via *urine and expired air within 21 days following administration.*

The compound, 2,3,7,8-tetrachlorodibenzo-*p*-dioxin (TCDD), is highly toxic (*see* Figure 1). The $LD_{50}$ for male and female rats given a single oral dose is 23 and 45 μg/kg, respectively, while that for male guinea pigs is 0.6 μg/kg (*1*). Recently, adverse effects were observed in a teratological study in which pregnant rats were given oral doses of 0.125–2.0 μg/kg/day TCDD from days 6 through 15 of gestation (*2*). The adverse effects were increased fetal mortality, early and late resorptions, and intestinal hemorrhage in the fetuses. No adverse effects were noted at the 0.03 μg/kg/day level.

In humans and rabbits contamination of the skin with TCDD produces chloracne-like lesions (*3, 4*). This disease is characterized by the appearance of hyperkeratosis, papules, comedones, and cysts.

There is no available information on the absorption, excretion, or tissue distribution of TCDD in animals. Therefore, this study was done to determine the excretion and tissue distribution of radioactivity derived from TCDD-$^{14}$C following a single oral dose of the labeled compound.

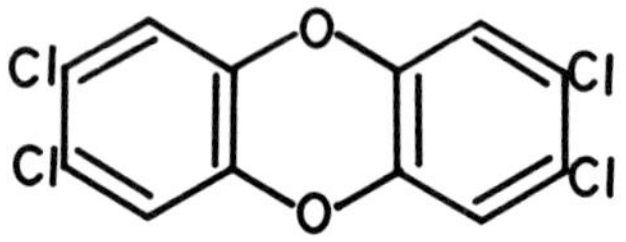

*Figure 1. Structural formula of 2,3,7,8-tetrachlorodibenzo-*p*-dioxin*

## *Methods*

**Animals.** Male Spartan strain Sprague Dawley rats weighing 165–210 grams were used. The rats were acclimated to the environment of the metabolism cages five days before dosage. Food and water were provided *ad libitum* throughout the experiment.

**$^{14}$C-Tetrachlorodibenzo-*p*-dioxin.** Uniformly labeled TCDD-$^{14}$C was synthesized at the Radiochemistry Research Laboratory of The Dow Chemical Co. The specific activity was 2.8 $\mu$Ci/mg. The TCDD-$^{14}$C sample was analyzed by mass spectrometry and gas–liquid chromatography and indicated a purity of 93.3 and 95.0%, respectively.

**Dosage.** TCDD was dissolved in acetone; subsequently, one part of the acetone solution was added to and mixed with nine parts of USP corn oil. The acetone–corn oil solution of TCDD was given to rats at a rate of 5 ml/kg by intubation. This volume of solution provided a dose of 50 $\mu$g/kg and 0.14 $\mu$Ci/kg TCDD-$^{14}$C.

**Sample Collection.** After administering the solution containing TCDD-$^{14}$C, the rats were placed in all glass metabolism chambers which were equipped for separate collection of urine, feces, and expired air. The $CO_2$ in the exiting air stream was trapped by bubbling it through 3:7 ethanolamine:2-methoxyethanol.

**Radioactivity Analysis.** Samples of urine, feces, and tissues were combusted to $^{14}CO_2$ and analyzed for radioactivity (5). By using this method the recovery of radioactivity from samples spiked with $^{14}$C was 95 $\pm$ 5%. To determine the radioactivity expired as $CO_2$, 5-ml aliquots of the solution used to trap the $CO_2$ were added to 15 ml of a scintillation counting solution containing 4 grams 2,5-diphenyloxazole (PPO) and 0.1 grams 1,4-bis-2(5-phenyloxazolyl)-benzene (POPOP) per liter of 1:1 toluene:2-methoxyethanol. Samples were counted for radioactivity in a Nuclear Chicago Mark II liquid scintillation counter. Counting efficiency was corrected by the internal standard technique.

## *Results*

The percentage of the total dose of radioactivity excreted daily in the feces, urine, or expired air over a 21-day period following a single oral dose of TCDD-$^{14}$C is shown in Figure 2. Approximately 30% of the

$^{14}C$ activity was excreted in the feces during the first 48 hours. Most of this probably represents unabsorbed TCDD-$^{14}C$. On each of the remaining 19 days, 1–2% per day of the $^{14}C$ activity was excreted in the feces. A total of 53.2 ± 3.8% of the administered dose was excreted *via* the feces over the 21-day period. The total cumulative amount excreted in the urine and expired air was 13.2 ± 1.3% and 3.2 ± 0.1%, respectively.

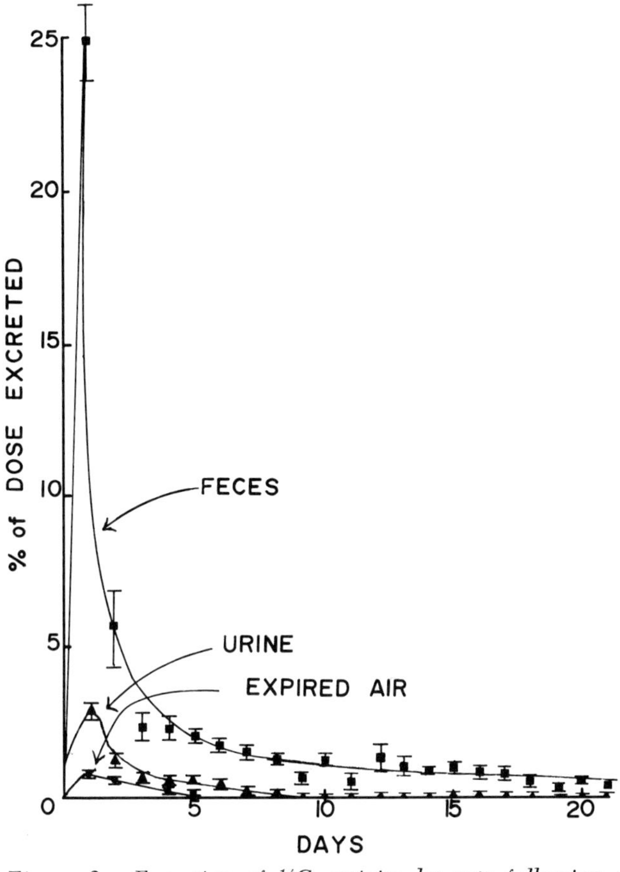


*Figure 2. Excretion of $^{14}C$ activity by rats following a single oral dose of 50 μg/kg (0.14 μCi/kg) 2,3,7,8-tetrachlorodibenzo-p-dioxin. Each point represents the mean ± SE for three rats.*

To determine the overall rate of clearance of $^{14}C$ administered as TCDD from the body, the total cumulative amount of $^{14}C$ excreted in feces, urine, and expired air at the end of each day was subtracted from the total dose administered to the animal. These values, representing the percentage of the total dose remaining in the animal at the end of each day, were then plotted semilogarithmically as a function of time

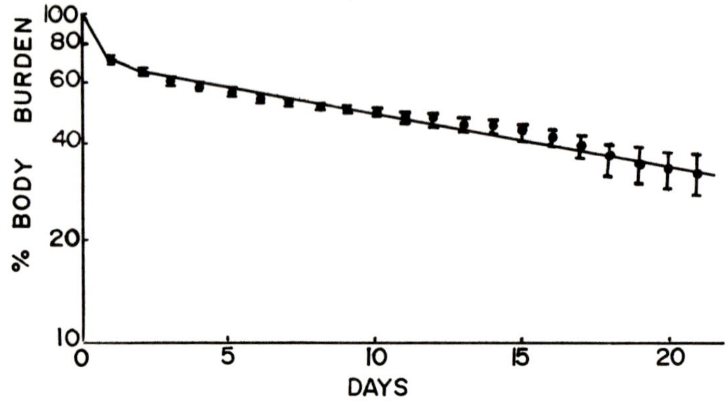


*Figure 3. Clearance of $^{14}C$ activity from the body of rats given a single oral dose of 50 μg/kg (0.14 μCi/kg) 2,3,7,8-tetrachlorodibenzo-p-dioxin. Each value represents the mean ± SE for three rats.*

(Figure 3). Except for the first two days following administration, the clearance of $^{14}C$ activity from the body followed apparent first order rate kinetics. The half-life for clearance, $t_{1/2}$, was 17.4 ± 5.6 days. As previously indicated, it was assumed that the relatively large amount excreted during the first 2 days had not been absorbed; therefore, these values were not used in calculating the clearance rate.

Analyses of tissues indicated that the $^{14}C$ activity derived from TCDD and/or its breakdown products was located mainly in the liver and fat (Table I). The percentages of the dose per gram of liver at 3, 7, and 21 days following administration were 3.18, 4.49, and 1.33, respectively. Comparable values for fat were 2.60, 3.22, and 0.43%/gram.

Smaller concentrations of $^{14}C$ activity were found in other tissues: muscle, testes, lungs, heart, skin, spleen, stomach, pancreas, brain, bone, kidneys, and adrenals (Table II). Standard errors as large as the mean

**Table I. $^{14}C$ Activity Expressed as Percent of Dose per Gram (%/gram) in the Liver and Fat of Rats 3, 7, and 21 Days Following a Single Oral Dose of TCDD-$^{14}C$[a]**

| | *Time Post-Administration* | | |
|---|---|---|---|
| *Tissue* | *3 days* | *7 days* | *21 days* |
| Liver | 3.18±0.21[b] (47%)[c] | 4.49±0.62 (45%) | 1.33±0.70 (11%) |
| Fat | 2.60±0.48 | 3.22±0.63 | 0.43[d] |

[a] Dose—50 μg/kg (0.14 μCi/kg); 3 rats/observation.
[b] Mean ± standard error.
[c] The percent of the total dose found in the entire liver.
[d] Mean for 2 rats.

**Table II. $^{14}C$ Activity Expressed as Percent of Dose per Gram (%/gram) in Various Tissues of Rats 3, 7, and 21 Days Following a Single Oral Dose of TCDD-$^{14}C$[a]**

| *Tissue* | *Time Post-Administration* 3 days | 7 days | 21 days |
|---|---|---|---|
| Muscle | 0.38±0.01 [b] | 0.21±0.05 | 0.20±0.12 |
| Testes | 0.38±0.03 | 0.36±0.10 | 0.11±0.09 |
| Lungs | 0.27±0.02 | 0.39±0.14 | 0.06±0.05 |
| Heart | 0.20±0.03 | 0.40±0.16 | 0.09±0.05 |
| Skin | 0.19±0.10 | 0.19±0.10 | 0.09±0.04 |
| Spleen | 0.15±0.02 | 0.95±0.53 | 0.22±0.22 |
| Stomach | 0.16±0.05 | 0.10±0.00 | 0.02±0.02 |
| Pancreas | 0.11±0.06 | 0.16[c] | 0.16±0.16 |
| Brain | 0.06±0.00 | 0.13±0.09 | 0.01±0.01 |
| Bone | 0.09±0.03 | 0.42±0.42 | 0.08±0.08 |
| Kidneys | 0.00 [d] | 0.34±0.17 | 0.00 [d] |
| Adrenals | 0.79±0.79 | 0.02±0.02 | 3.69±1.77[e] |

[a] Dose—50 μg/kg (0.14 μCi/kg); 3 rats/observation.
[b] Mean ± standard error.
[c] Mean of 2 rats.
[d] No activity above background in all 3 rats.
[e] This large value may be experimental error. The DPM's above background for the adrenals of the three rats were 17, 29 and 90. Since the total amount of tissue was less than 20 mg, the multiplication factor may have magnified the error many fold.

suggest that some of the values presented in this table may result from experimental error: adrenals—3 days; bone—7 days; spleen—21 days; pancreas—21 days. Radioactivity exceeding background in these tissues at the indicated time was detected in only one of three rats.

The value given in Table II for adrenals 21 days following administration suggests that this tissue may concentrate TCDD-$^{14}C$ and/or a metabolite. This observation results probably from experimental error. The disintegrations per minute (DPM) above background for this tissue were only 17, 29, and 90. Since the total amount of tissue available for analysis was less than 20 mg, the multiplication factor may have magnified the error many fold.

Total recovery of the administered $^{14}C$ activity was determined for those rats used in the 21-day experiment. The $^{14}C$ activity remaining in the unused carcass was determined by analyzing an aliquot of a homogenate of the remaining carcass. The recovery was 96.8 ± 3.0%.

## *Discussion*

In this study the tissue distribution and excretion of $^{14}C$ activity has been evaluated in rats following a single oral dose of TCDD-$^{14}C$. Almost 30% of the dose administered was eliminated *via* the feces during the first 48 hours following treatment. The excretion of $^{14}C$ activity *via* the feces after the first 48 hours ranged from 1–2% per day. It appears that TCDD is incompletely absorbed from the gastrointestinal tract. The $^{14}C$ activity derived from the absorbed TCDD-$^{14}C$ also is excreted mainly *via* the feces.

Once absorbed in the body, most of the $^{14}C$ activity derived from TCDD-$^{14}C$ is localized in the liver and fat. The data suggest that the level in these tissues is approximately 10-fold greater than that found in other tissues. The $^{14}C$ level in liver and fat seemed to increase between 3 and 7 days following administration; the $^{14}C$ activity in liver and fat decreased more between 7 and 21 days than what would have been predicted by assuming that the rate of clearance from these tissues would be equal to the rate of clearance from the body. Between days 7 and 21 the $^{14}C$ level in muscle remained essentially unchanged. Therefore, redistribution of TCDD or metabolites of TCDD may have been occurring. The apparently high level in the adrenals 21 days after administration results probably from experimental error.

The dose of TCDD given to the male rats used in this study, 50 $\mu g/kg$, was approximately twice the $LD_{50}$, 23 $\mu g/kg$. This large dose was necessary because of the low specific activity of the TCDD-$^{14}C$ used. In this study rats lost weight, and their physical condition was poor, which typifies the insidious response to TCDD (*1*). Survival of the rats for 21 days was not totally unexpected because in previous studies on the lethality of TCDD deaths frequently occurred 20 or more days following a single oral dose of similar magnitude (*1*). With doses that do not induce untoward effects, the compound may be excreted at a different rate.

The results do not differentiate between $^{14}C$ activity derived from TCDD and that of possible metabolites. However, small amounts of $^{14}C$ activity were detected in the expired air and urine within the first 10 days following administration. This is evidence that some metabolic alteration or breakdown of TCDD occurs.

## Acknowledgment

The authors thank W. W. Muelder for preparation of $^{14}C$-2,3,7,8-tetrachlorodibenzo-*p*-dioxin.

## Literature Cited

1. Rowe, V. K., Norris, J. M., Sparschu, G. L., Schwetz, B. A., Gehring, P. J., ADVAN. CHEM. SER. (1973) **120**, 55.
2. Sparschu, G. L., Dunn, F. L., Rowe, V. K., "Study of the Teratogenicity of 2,3,7,8-Tetrachlorodibenzo-*p*-dioxin in the Rat," *Food Cosmet. Toxicol.* (1971) **9**, 405.
3. Kimmig, J., Schulz, K. H., *Dermatologia* (1957) **115**, 540.

4. Jones, E. L., Krizek, H., "A Technique for Testing Acnegenic Potency in Rabbits, Applied to the Potent Acnegen, 2,3,7,8-Tetrachlorodibenzo-*p*-dioxin," *J. Invest. Dermatol.* (1962) **39,** 511.
5. Smith, G. N., Ludwig, P. D., Wright, K. C., Bauriedel, W. R., "Simple Apparatus for Combustion of Samples Containing $C^{14}$-Labeled Pesticides for Residue Analyses," *Agr. Food Chem.* (1964) **12,** 172.

RECEIVED February 8, 1972.

# 10

# An Improved Analysis for Tetrachlorodibenzo-*p*-dioxins

ROBERT BAUGHMAN and MATTHEW MESELSON

Department of Chemistry and Department of Biochemistry and Molecular Biology, Harvard University, Cambridge, Mass. 02138

*A meaningful assessment of the environmental levels of 2,3,7,8-tetrachlorodibenzo-*p*-dioxin (TCDD), an extraordinarily toxic compound present as an impurity in the herbicide 2,4,5-T and in some commercial chlorophenols, can be made only by evaluating representative samples with a sufficiently sensitive analytical method. The sensitivity required is well beyond that available with current methods. We report a procedure using time averaged high resolution mass spectroscopy with a sensitivity ($10^{-12}$ gram) suitable for such an investigation. Interference from pentachlorobiphenyl in certain materials from the environment presently limits attaining full sensitivity of the method although we are working toward a resolution of this problem.*

Our interest in the chlorodioxin problem stems from our work with the Herbicide Assessment Commission of the American Association for the Advancement of Science which was organized in 1970 to initiate a study of the effects of herbicide use in Vietnam. As one part of that investigation we are analyzing various samples from Vietnam for TCDD, a known impurity in 2,4,5-T (*1, 2, 3, 4*). This herbicide in a one-to-one mixture with 2,4-D is a component of agent Orange, the herbicide that was used most widely in Vietnam. Our aim has been to determine whether TCDD has accumulated in food chains to any significant extent.

We were surprised to find that no method existed that was sensitive enough to detect TCDD in animal tissues even after administration in some species of lethal doses. An example is the guinea pig, the most susceptible species of the few that have been tested, and therefore a good choice for establishing desirable limits of detection. The lethal single oral dose ($LD_{50}$) in males of this species is 0.6 $\mu$g/kg body weight

(5). This means that if all of the TCDD were retained, the level of TCDD would be less than 1 part per billion (ppb) in the whole animal. The lowest reported limit of detection for TCDD in whole tissue is 50 ppb (6). Thus, a guinea pig could be killed with TCDD, and it would be impossible to establish this fact with the analytical procedures in current use.

Such analytical procedures are clearly of little value in monitoring food chains for the buildup of TCDD. This is even more apparent if one considers the possibility of sub-lethal toxic effects and allows a margin for a safety factor. If we provide a factor of about 100 for non-lethal toxicity (6) and a further factor of 10 for a safety margin and to allow for the possible existence of species even more sensitive than the guinea pig, we would require a level of detection of $(10^{-9})(10^{-2})(10^{-1}) = 10^{-12}$ or 1 ppt (1 part in $10^{12}$) for environmental monitoring. For a 1 gram sample this would require a limit of detection of 1 pg ($10^{-12}$ gram). The limit of detection of TCDD for the electron capture detector, the keystone of current analytical procedures, is not much less than 1 ng ($10^{-9}$ gram).

In addition to high sensitivity, a requirement for any acceptable analytical method is high specificity because at very low levels few confirmatory procedures can be used to establish the identity of a particular compound. A method which uniquely combines high sensitivity with high specificity is high resolution mass spectrometry. We have used this method as the basis for an approach which we believe will make possible a meaningful assessment of TCDD levels in the environment.

Figure 1 shows that the mass spectrum of TCDD is relatively simple. (All the work reported here was done with an Associated Electrical In-

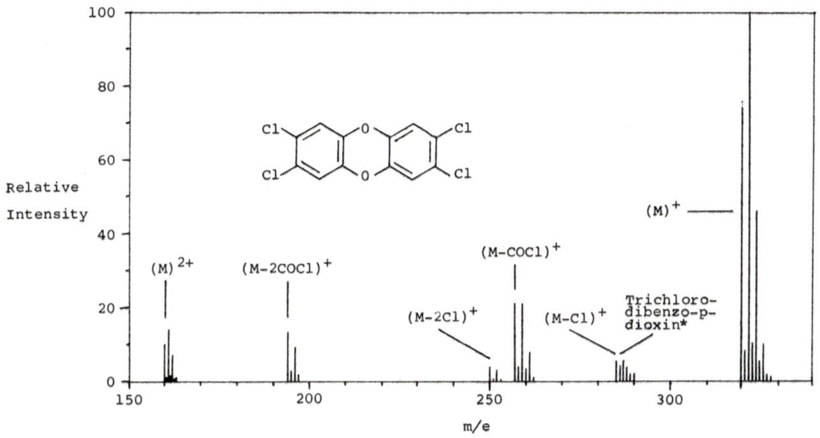


*Figure 1. Mass spectrum of 2,3,7,8-tetrachlorodibenzo-p-dioxin (TCDD). The molecular ion ($M^+$) is at m/e 320. Ionizing voltage 70 eV, source 150°C. Asterisk denotes impurity.*

dustries MS-9 double focusing mass spectrometer.) The base peak is the molecular ion at *m/e* 320. As a result of the various possible combinations of the naturally occurring $^{35}Cl$ and $^{37}Cl$ isotopes in a tetrachloro compound, the signal for the molecular ion is a pentuplet with peaks at *m/e* 320, 322, 324, 326, and 328 with intensities in the ratio 77:100:49:10:1. In addition, the four chlorine atoms and the limited number of hydrogen atoms make the compound significantly mass deficient (the *m/e* 320 peak is actually 319.8956) and, therefore, relatively easily resolved from most other organic residues (for which *m/e* 320 would be 320.1–320.2).

First, we tried scanning the region *m/e* 310–330 (Figure 2). As the sample was introduced into the mass spectrometer, signals appeared at *m/e* 320, 322, and 324 and then, as the sample became exhausted, disappeared. Under these conditions the limit of sensitivity was on the order of 100 pg. We next reduced the scanning interval to about one third of a mass unit. This allows the detector to spend more time in the region of interest, considerably increasing the signal. At a resolution of 10,000 a series of scans was made, alternately two at 322, two at 314 perfluorotributylamine (PFA) reference peak, two at 322, etc. (Figure 3). The PFA was bled in from an external reservoir at a constant rate, providing reference peaks that remain at the same height throughout the analysis while the sample peaks rise and then fall as the sample volatilizes. This procedure with a sensitivity of about 20 pg was still not adequate.

It is possible to obtain greater sensitivity from the repeated narrow scans shown in Figure 3 by combining them to produce a single time averaged scan. Procedures accomplishing this under low resolution conditions have been reported previously (*7, 8*). Under the present conditions a system was devised for doing this using a Varian 1024 averaging computer (CAT) in conjunction with the MS–9. The result is shown in Figure 4. The signal for a pair of peaks at the limit of detection for a single scan is shown in Figure 4A, and the averaged signal from sixty scans is shown in Figure 4B. The signal-to-noise ratio is expected to improve approximately as the square root of the number of scans (*9*). With 1 min of scanning at a rate of one scan per second, the observed improvement is approximately that expected. At very fast scan rates data is inefficiently transferred to the memory of the CAT, and resolution is decreased by damping caused by the time constant of the MS–9 circuitry. In the present system this limits the maximum scan rate to four scans per second. With very short volatilization times ($<$ 10 sec) sensitivity is decreased, perhaps in part because of decreased ionization efficiency. With volatilization times longer than about 60 sec the drift in peak position from scan to scan is large enough to decrease significantly the resolution observed in the time averaged spectrum. The optimum volatilization time is from 30 to 60 sec.

The interfacing of the CAT with the MS–9 is illustrated in Figure 5. The ions in the *m/e* region of interest, after being focussed, pass by a small magnet coil which deflects the beam back and forth over the detector slit. After passing through the slit, the ions strike an electron

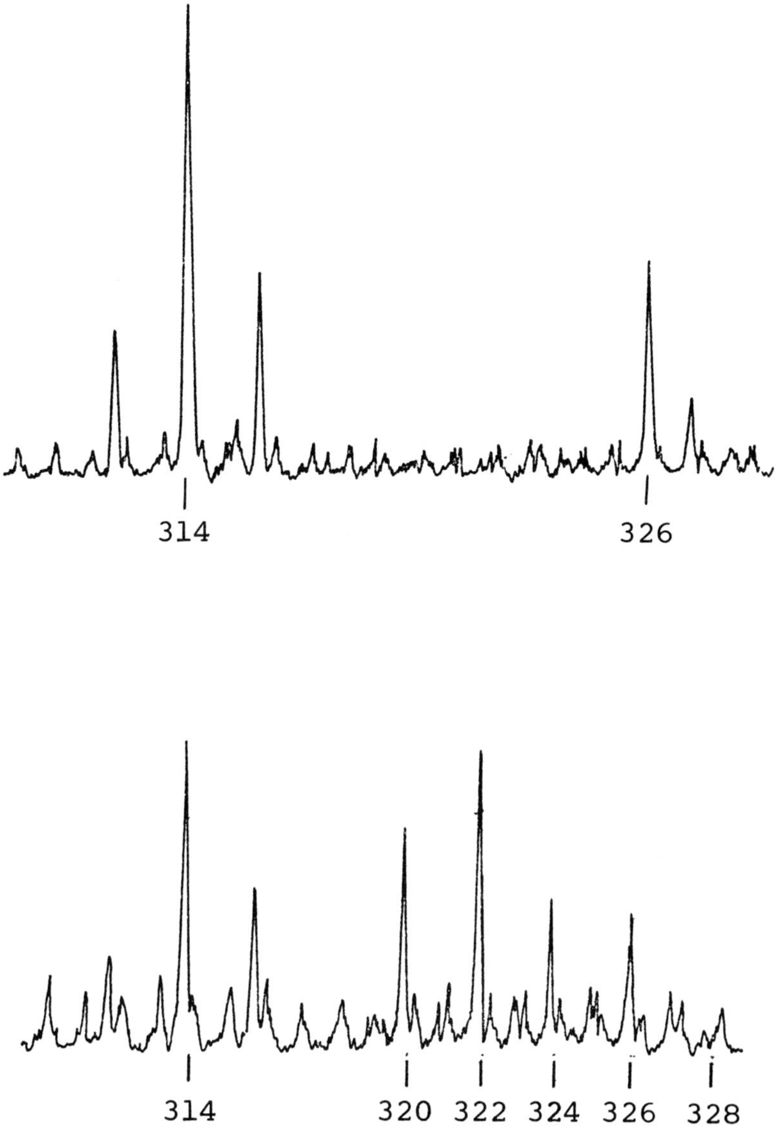


*Figure 2. Repetitive scanning of* m/e *310–330 (5 sec/scan). Standard conditions for this and all following figures: ionizing voltage 70 eV, accelerating voltage 8 kV, trap current 300 μA, multiplier 600, source 150°C.*

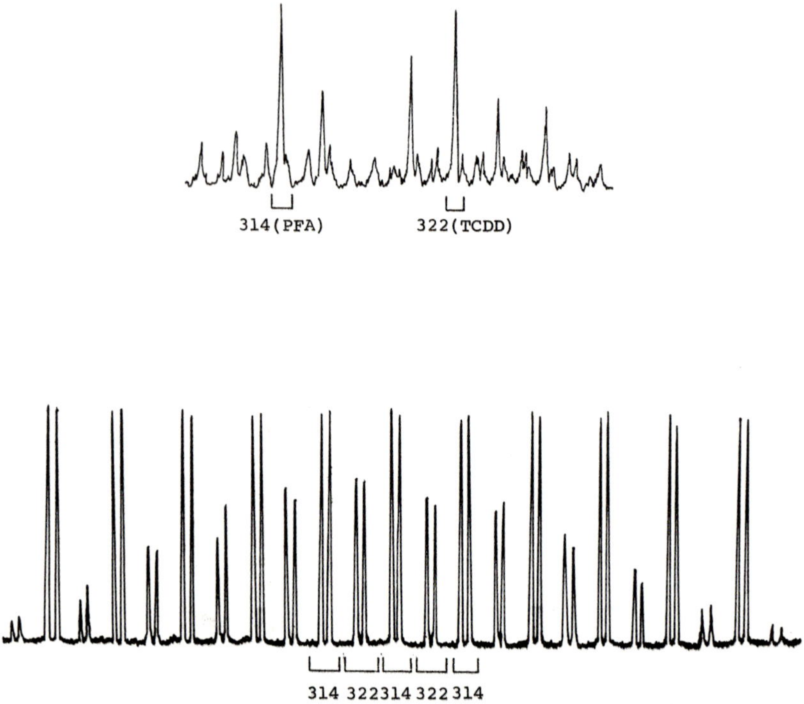


*Figure 3. Alternating narrow scans*

*Top: regions scanned on each narrow scan. Bottom: alternating narrow scans (two half-second scans at 314, two at 322, etc.) (200 pg TCDD).*

multiplier, producing a signal which is continuously displayed on an oscilloscope on the MS–9. This provides a means of monitoring each scan. Simultaneously, the signal is added to the 1024-channel memory of the CAT. An oscilloscope on the CAT continuously displays the total memory content which makes it possible to monitor the overall course of the analysis. A potential problem of phasing the beam deflection coil with the memory sweep circuit of the CAT is avoided by using the sweep voltage ramp of the CAT, *via* an amplifier and appropriate circuits of the MS–9, to drive the beam deflection coil. The coil is thus necessarily in synchrony with the CAT.

The procedure we have adopted for introducing samples into the MS–9 is shown in Figure 6. It provides reproducible analyses at a high level of sensitivity. The sample tubes are made from 1 mm id melting point capillaries. A Hamilton 10-$\mu$l syringe is used to introduce a 3–4 $\mu$l portion of the residue into the sample tube. With a small flame the sample tube is drawn out just above the level of the liquid to produce a capillary constriction about 20 mm long. The solvent is then removed at reduced pressure. Bumping is prevented by the capillary constriction. The sample tube is then sealed with a flame. At the time of analysis the capillary is broken off 2–3 mm above the constriction to give the tube configuration shown in Figure 6. The tubes are introduced into the MS–9 with a wire holder on the tip of a standard MS–9 direct insertion probe. To aid reproducibility all analyses are started at the same time after insertion of the sample tube into the MS–9 source. The temperature of the source heating block is adjusted to give a sample volatilization time of 30 to 60 seconds.

The result of combining these various components in the analysis of a 2-pg sample of TCDD is illustrated in Figure 7. An internal standard is given by a PFA fragmentation peak which is a known distance, 85 mmu (1 millimass unit or mmu = $10^{-3}$ atomic mass unit), from the TCDD peak. In its present form the MS–9–CAT system has a limit of detection for TCDD of about 1 pg.

The procedure we have described retains the generality of normal mass spectral analysis. It is particularly suited, however, to compounds

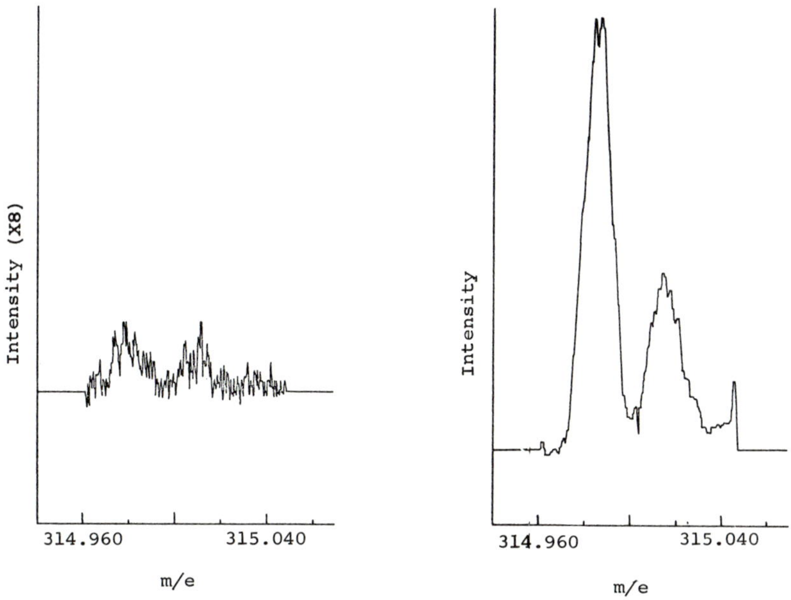


*Figure 4. Improvement in sensitivity with the CAT. PFA and a reference peak at* m/e *315. The observed improvement in signal-to-noise ratio results from the longer total scanning time and also the fact that many sweeps are made during this time. The overall improvement in signal-to-noise ratio depends on the detailed power spectrum of the noise* (9). *Resolution 12,000 here and for all following time averaged spectra*

*Left: one scan. Right: 60 scans (one scan/sec).*

containing atoms with significant mass defects, such as heavy metal or organochlorine compounds, which are easily resolved from other residues.

Before the procedure is applied to tissue or other samples from the environment, some potential complications must be taken into account. One is the possibility that other chlorinated organic compounds present in the environment might interfere with or obscure the TCDD peaks. To test this, we obtained mass spectra on the MS–9 for most of the common organochlorine pesticides including lindane, aldrin, dieldrin, mirex, heptachlor, DDD, DDE, and DDT, as well as various polychlorinated biphenyl (PCB) mixtures. In the TCDD mass range DDE from its molecular ion has isotopic isomer peaks at $m/e$ 320, 322, and weakly 324. The molecular ion of pentachlorobiphenyl, a component of some PCB mixtures, has a peak at $m/e$ 324, and this compound has a weak fragmentation peak at $m/e$ 322. DDT has weak fragmentation peaks at $m/e$ 320, 322, and 324. As shown for $m/e$ 322 in Figure 8, all of these compounds can be resolved from TCDD at our normal resolution of 12,000 (27 mmu at $m/e$ 322). The relative input amounts of each compound producing the peaks shown are: DDT, 250; DDE, 25; TCDD, 1; PCB (Arochlor 1254), 250. Even though moderately large excesses of these interferences can be tolerated, it is necessary to use highly

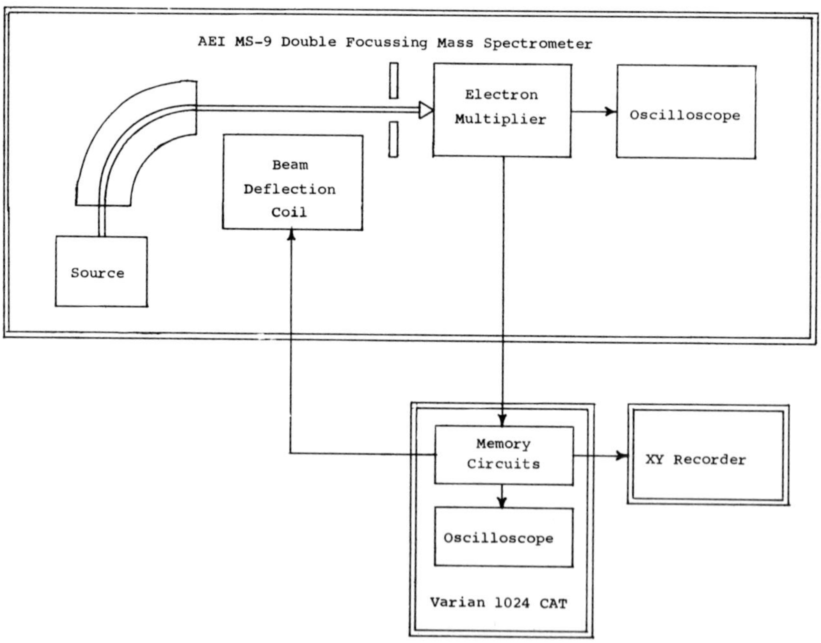


*Figure 5. CAT–MS-9 interfacing*

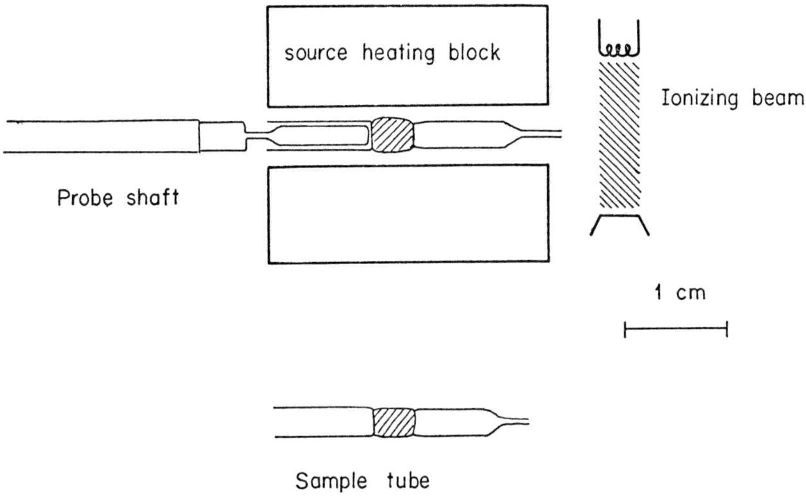


*Figure 6. Sample introduction*

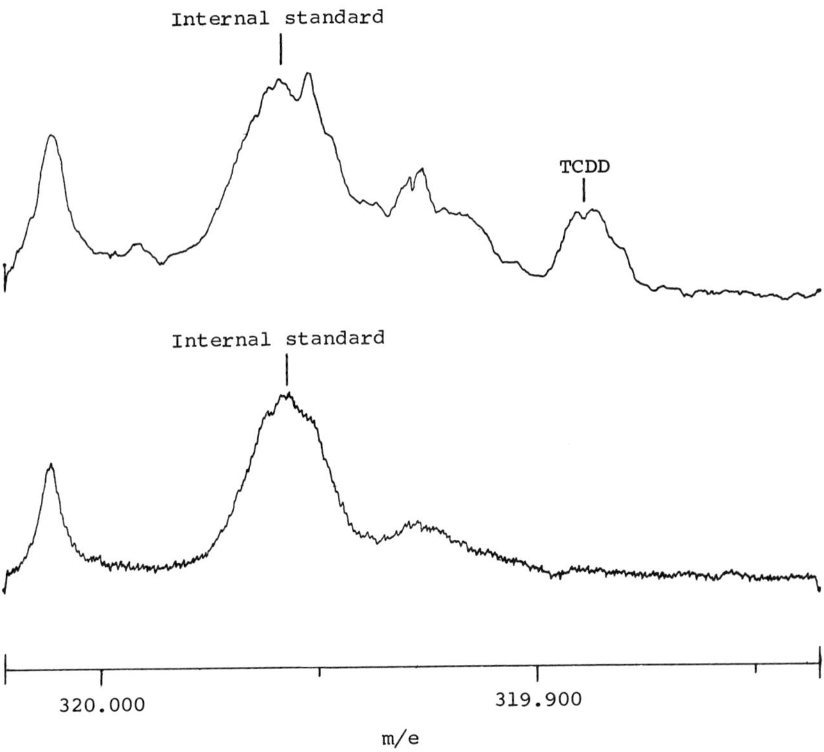


*Figure 7. Limit of detection*

*Top: 2 pg ($2 \times 10^{-12}$ gram) TCDD. Bottom: background.*

efficient cleanup procedures to carry out analyses for TCDD at very low levels.

Another complicating characteristic of materials from the environment is that the size and nature of the residue to be analyzed in the mass spectrometer will change from sample to sample. To determine if this might have an effect on the observed TCDD signal, we analyzed identical samples of TCDD with differing amounts of squalane, a saturated hydrocarbon selected as a model for residues obtained from standard extraction and cleanup procedures. As is indicated in Table I (Part A), there was

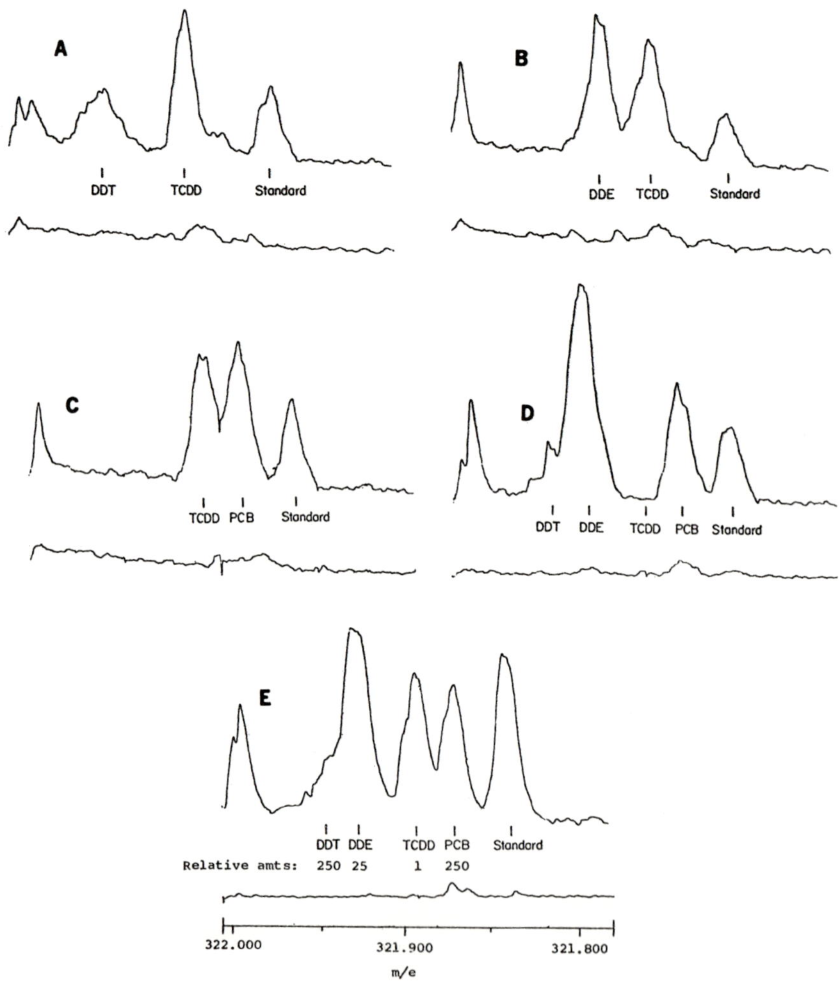


*Figure 8. Resolution of TCDD from interferences*

*A: DDT + TCDD. B: DDE + TCDD. C: PCB (Arochlor 1254) + TCDD. D: DDT + DDE + PCB. E: DDT + DDE + PCB + TCDD.*

**Table I. Effect of Size of Total Residue**

*A. Response for TCDD*

| *Squalane added (micrograms)* | *Relative response for 20 pg TCDD* |
|---|---|
| 0 | 100 |
| 1 | 75 |
| 5 | 50 |
| 25 | 15 |

*B. Ratio of two components* [a]

| *Squalane added (micrograms)* | *TCDD/TBB* |
|---|---|
| 1 | 11 |
| 3 | 5 |
| 9 | 1 |

[a] Ratio of response for 20 pg TCDD to response for 200 pg of 2,3,5,6-tetrachloro-4-bromoethylbenzene (TBB).

a significant effect on the response to TCDD. This eliminates any hope of determining the amount of TCDD present directly from the area of the TCDD peak. In an attempt to resolve this difficulty, we added an internal standard which does not coincide with TCDD or with peaks from any of the potential chlorinated hydrocarbon impurities and measured the ratio between TCDD and the standard in the presence of various amounts of squalane. A compound with a suitable mass, 2,3,5,6-tetrachloro-4-bromoethylbenzene (TBB), was synthesized for this purpose. However, as is apparent from Table I (Part B), the ratio of its signal to that of TCDD was far from constant, and this procedure was ruled out. Table I also suggests that in order not to cause a significant loss of sensitivity, the total sample size should be kept under 5 $\mu$g.

Some alternative method had to be devised to quantify the TCDD measurements. The problem was solved with the observation, illustrated in Figure 9, that the response to TCDD is linear over a wide concentration range as long as the size and nature of the sample matrix remain the same. Thus, it is possible to divide a sample into two equal portions, run one, then add an appropriate known amount of TCDD to the other, run it, and by simply noting the increase in area caused by the added TCDD to calculate the amount of TCDD present in the first portion. Figure 9 illustrates the reproducibility of the system. Each point was obtained from four or five independent analyses with an error (root mean square) of 5–10%, as indicated by the error flags, which is acceptable for the present purposes.

To satisfy the requirement of having a total residue of only a few micrograms, the sample cleanup must be very thorough (*10*). The procedure which we have used to accomplish this is the following. A 10-

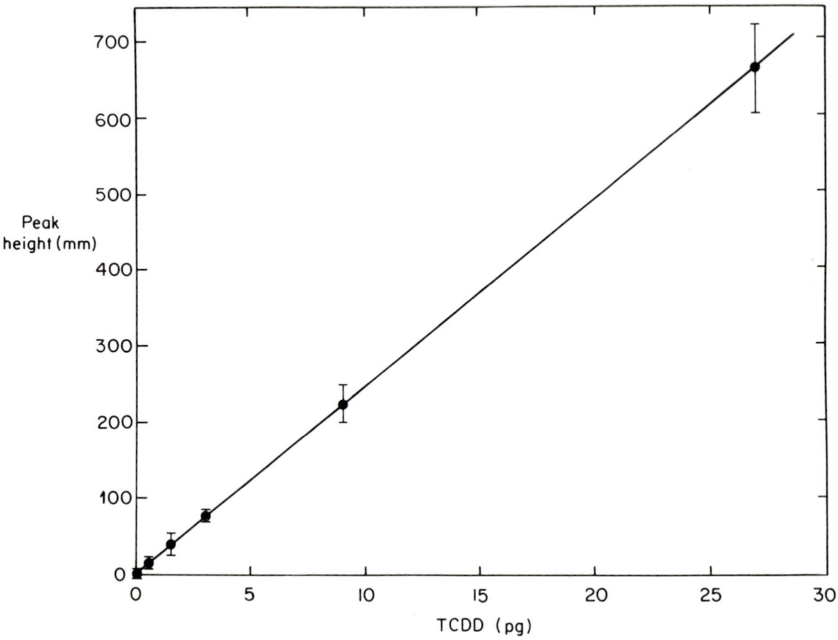


*Figure 9. Linearity of response and reproducibility. The error flags indicate the root mean square error for five measurements at each value. The average relative error is about 10%.*

gram sample of human milk was combined with 10 ml of ethanol (all solvents were of pesticide grade) and 20 ml of aqueous 40% KOH and refluxed for 6 hours. This solution was then extracted with four 6-ml portions of 5% benzene in hexane. The organic phase was extracted with four 8-ml portions of 85% $H_2SO_4$, filtered through a 10 mm id column of 10 grams of powdered $Na_2CO_3$, concentrated carefully to about 10 $\mu$l, and preparatively chromatographed by gas–liquid chromatography on a 1 m × 1/4″ column of 3% OV–101 methyl silicone polymer liquid phase on 50/60 mesh Anakrom AB solid support (200°C column, 30 ml/min He). The GLC trap consisted of a 150 mm × 1.5 mm id borosilicate glass tube packed with 30 mm of 100/120 mesh glass beads retained with glass wool plugs (*11, 12*). The trap was wetted with hexane and cooled in dry ice. The GLC cleanup is carried out through a thermal conductivity detector. A small amount of an internal standard, *m*-terphenyl, with a known retention time relative to TCDD was added to make certain that the TCDD collection was carried out at the right retention time. The total residue from the GLC cleanup when divided into twelve fractions provided a suitable sample size.

Figure 10 illustrates the results of a typical analysis of a 10-gram sample of human milk to which 0.1 ppb TCDD has been added. Blank

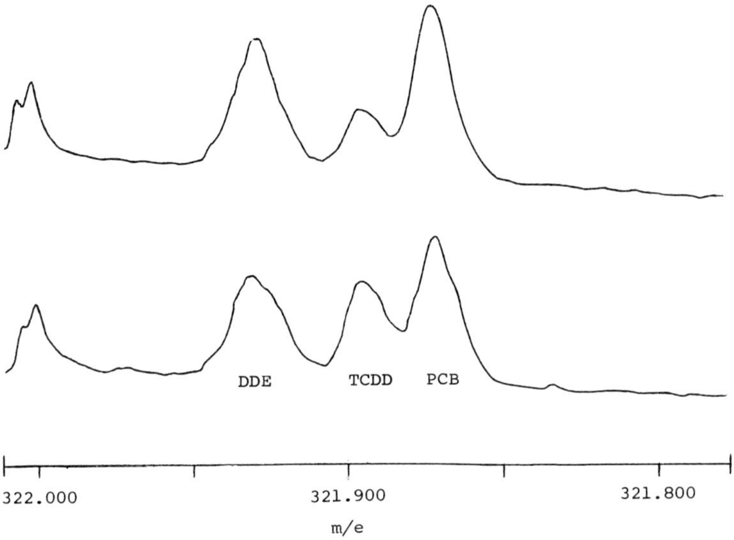


*Figure 10. TCDD recovery in 10 gram human milk to which $10^{-9}$ gram (0.1 ppb) TCDD was added. Each trace represents 8% of the total residue. Top: residue (25% recovery). Bottom: residue + 20 pg TCDD.*

samples showed no indication of TCDD at this level. Each trace represents 8% of the total sample and since the area under the unknown TCDD peak, calculated from the run with TCDD added, corresponds to approximately 20 pg, the recovery is about 25%. The other signals observed in the TCDD mass region are probably from DDE and pentachlorobiphenyl. At the present time the major impediment to extending the procedure to the desired level of 1 ppt is interference from pentachlorobiphenyl. We are investigating changes in the cleanup procedure which show excellent promise of reducing the level of this and other interferences in the final residue. We are also developing a procedure for directly measuring the recovery of TCDD in each cleanup by adding chlorine–[$^{37}$Cl] labelled TCDD to the sample before cleanup.

### *Acknowledgment*

We would like to thank David Firestone of the Division of Chemistry and Physics, FDA for providing us with samples of TCDD, Klaus Biemann and Charles Hignite of the Department of Chemistry, Massachusetts Institute of Technology for assistance in the early stages of this work, David Parrish of the Department of Chemistry, Harvard University for assistance in developing the MS–9–CAT system, and William Doering of the Department of Chemistry, Harvard University for the use of laboratory facilities. This work was supported by the Herbicide Assessment Commission of the American Association for the Advancement of Science and by the Ford Foundation.

### *Literature Cited*

1. Brenner, K. S., Müller, K., Sattel, P., *J. Chromatogr.* (1972) **64,** 39.
2. Elvidge, D. A., *Analyst* (1971) **96,** 721.
3. Johnson, J. E., Hearings Before the Subcommittee on Energy, Natural Resources, and the Environment of the Committee on Commerce, United States Senate on "Effects or 2,4,5-T on Man and the Environment," April 7 and 15, 1970.
4. Woolson, E. A., Thomas, R. F., Ensor, P. D. J., *J. Agr. Food Chem.* (1972) **20,** 351.
5. "Report on 2,4,5-T," A Report of the Panel on Herbicides of the President's Science Advisory Committee, Executive Office of the President, Office of Science and Technology (March 1971).
6. Woolson, E. A., Reichel, W. L., Young, A. L., ADVAN. CHEM. SER. (1973) **120,** 112.
7. Biros, F. J., *Anal. Chem.* (1970) **42,** 537.
8. Plattner, J. R., Markey, S. P., *Org. Mass Spec.* (1971) **5,** 463.
9. Ernst, R. R., *Rev. Sci. Inst.* (1965) **36,** 1689.
10. Ress, J., Higginbotham, G. R., Firestone, D., *J. Ass. Offic. Anal. Chem.* (1970) **53,** 628.
11. Bierl, B. A., Beroza, and Ruth, J. M., *J. Gas Chromatogr.* (1968) **6,** 286.
12. Murray, K. E., Shipton, J., Robertson, A. V., Smyth, M. P., *Chem. Ind.* (1971) 401.

RECEIVED February 8, 1972.

# 11

# Environmental Significance of Chlorodioxins

PHILIP C. KEARNEY, ALLAN R. ISENSEE, CHARLES S. HELLING, EDWIN A. WOOLSON, and JACK R. PLIMMER

Agricultural Environmental Quality Institute, Agricultural Research Service, U. S. Department of Agriculture, Beltsville, Md. 20705

*An environmental protocol has been developed to assess the significance of newly discovered hazardous substances that might enter soil, water, and the food chain. Using established laboratory procedures and $^{14}$C-labeled 2,3,7,8-tetrachlorodibenzo-*p*-dioxin (TCDD), gas chromatography, and mass spectrometry, we determined mobility of TCDD by soil TLC in five soils, rate and amount of plant uptake in oats and soybeans, photodecomposition rate and nature of the products, persistence in two soils at 1, 10, and 100 ppm, and metabolism rate in soils. We found that TCDD is immobile in soils, not readily taken up by plants, subject to photodecomposition, persistent in soils, and slowly degraded in soils to polar metabolites. Subsequent studies revealed that the environmental contamination by TCDD is extremely small and not detectable in biological samples.*

Detection of the highly potent impurity, 2,3,7,8-tetrachlorodibenzo-*p*-dioxin (TCDD) in the herbicide 2,4,5-trichlorophenoxyacetic acid (2,4,5-T), necessitated an environmental assessment of the impact of this contaminate. Information was rapidly needed on movement, persistence, and plant uptake to determine whether low concentrations reaching plants, soils, and water posed any threat to man and his environment. Because of the extreme toxicity of TCDD, utmost precautions were taken to reduce or minimize the risk of exposure to laboratory personnel. Synthesis of uniformly labeled $^{14}$C-TCDD by Muelder and Shadoff (*1*) greatly facilitated TCDD detection in soil and plant experiments. For unlabeled experiments it seemed wise to use only small quantities of diluted solutions in situations where decontamination was feasible and to rely on the sensitivity afforded by electron capture gas chromatography as reported by Woolson *et al.* (2). In addition to movement, persistence, and plant uptake studies, it was essential to determine whether chlorinated phenols could be condensed to their corresponding chlorodioxins by photochemical or biochemical processes under natural conditions.

**Table I. Properties of Soils Used in Mobility Studies with the Chlorodioxins**

| *Soil Type* | *Organic Matter, %* | *Clay, %* | *Field Moisture Capacity, %* | *pH* |
|---|---|---|---|---|
| Norfolk sandy loam | 0.14 | 11.3 | 6.5 | 5.1 |
| Lakeland sandy loam | 0.90 | 12.0 | 8.5 | 6.4 |
| Hagerstown silty clay loam | 2.50 | 39.5 | 25.8 | 6.8 |
| Barnes clay loam | 6.90 | 34.4 | 28.5 | 7.4 |
| Celeryville muck | 90.4 | | 113 | 5.0 |

### *Mobility*

The mobilities of two chlorinated dioxins, 2,7-dichlorodibenzo-*p*-dioxin (DCDD) and TCDD, and 2,4,5-T were evaluated by the soil thin layer chromatography (TLC) technique developed by Helling and Turner (3). This is a simple, safe technique (*i.e.*, it is easy to decontaminate), that gives comparative data for other chemical compounds analyzed in the same system. The procedure uses the conventional techniques used in preparing silica gel or alumina thin layer plates. Instead of silica gel, or alumina, however, sieved soils are used as the solid support. Radioisotopically-labeled compounds are spotted at the origin, and the plate is developed with water, dried, and covered with no screen x-ray film to locate the compound. Five soils varying in texture and other properties were tested: Norfolk and Lakeland sandy loams, Hagerstown silty clay loam, Barnes clay loam, and Celeryville muck (Table I). These soils gave a cross section of important soil parameters affecting mobility. Both dioxins were immobile in all soils—*i.e.*, they would not be leached into the soil by rainfall or irrigation (4).

The relative immobility of the chlorodioxins is expected, based on the very low solubility of these compounds in water (0.6 $\mu$g/liter). In contrast, the herbicide, 2,4,5-T, is relatively mobile in sandy soils, but movement decreases as soil organic matter increases. What does this information tell us, and how does it compare with other organic compounds? A mobility scale has been devised for a large number of pesticides (3). Higher mobility numbers reflect increased compound mobility in soils. The dioxins would be in Class 1—*i.e.*, they are immobile in soils and would compare with several chlorinated hydrocarbon insecticides.

Thus, the dioxins would not threaten groundwater supplies and would be subject to the surface processes affecting pesticides.

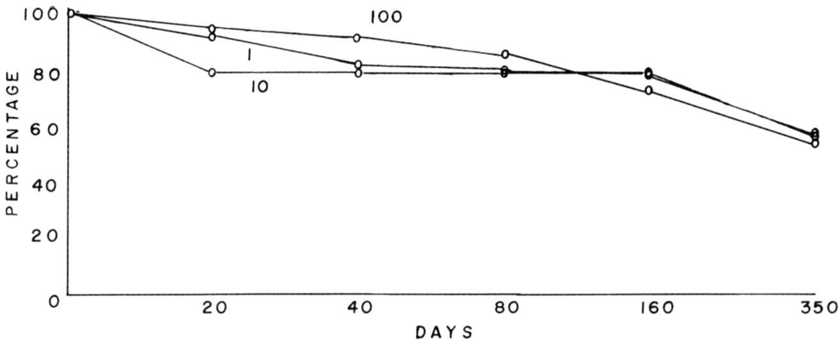


*Figure 1. Persistence of TCDD in Lakeland sandy loam applied at 1, 10, 100 ppm and sampled after 20, 40, 80, 160, and 350 days*

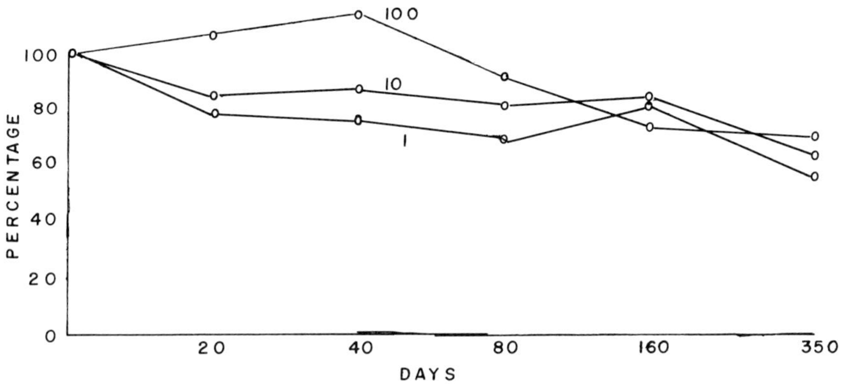


*Figure 2. Persistence of TCDD in Hagerstown silty clay loam at 1, 10, 100 ppm and sampled after 20, 40, 80, 160, and 350 days*

### *Persistence*

The objectives of the soil persistence experiments were (1) to learn the effect of soil type and concentration on the TCDD degradation rate, (2) to isolate and characterize degradation products from DCDD and TCDD, and (3) to determine whether chlorodioxins could be formed from chlorophenol condensation in the soil environment. This last study was essential since quality control at the manufacturing level could reduce or eliminate the formed dioxin impurity. But the biosynthesis of chlorodioxins by chlorophenol condensation in the soil environment could not be controlled and would have connotations for all chlorophenol-derived pesticides if formation did occur. The same question needed to be answered for photochemical condensation reactions leading to chlorodioxin formation. No evidence of DCDD or TCDD could be found in soils receiving 10, 100, and 1000 ppm of the 2,4-dichlorophenol or the 2,4,5-trichlorophenol, as determined by electron capture gas chromatography. Plimmer and Klingebiel (5) found no evidence for photocondensation of lower chlorophenols to chlorodioxins. Therefore, the possibility of chlorodioxins arising as biochemical or photochemical processes in the environment seems remote. These findings were critical for assessing the continued use of chlorophenolic pesticides or any chlorophenol used on a large scale basis.

To determine the persistence of chlorodioxins, concentration rates of 1, 10, and 100 ppm of unlabeled TCDD were established in 300-gram soil samples and then assayed periodically for dioxin residues. Several factors were considered in selecting these rates—they are many times greater in magnitude than would be encountered in the field. First, since the stability of these dioxins in soils was unknown, three concentrations were used to determine the degradation pattern, particularly if the process were rapid. Second, the highest concentration was to be used for product identification studies. Finally, preliminary problems in detection necessitated a range of concentrations exceeding 1 ppm since the limit of detection was *ca.* 0.2 ppm under the conditions used.

Measurements on dioxin residues after 20, 40, 80, 160, and 350 days of incubation of 28 $\pm$ 3°C in foil sealed beakers indicated a relatively slow degradation process in both soils (Figures 1 and 2). At 350 days, 56% of the initially applied TCDD was recovered from the sandy soil and 63% from the silty clay loam for all concentrations. A second experiment was conducted using $^{14}$C-TCDD at 1.78, 3.56, and 17.8 ppm in the same two soils. After 350 days the soils were combusted, and $^{14}CO_2$ trapped and determined by liquid scintillation. The amount recovered from Lakeland was 67% at 1.78 ppm, 70% at 3.56, and 73% at 17.8 ppm. For Hagerstown the recoveries were 52%, 52%, and 89% over the same concentrations. The $^{14}$C and gas–liquid chromatography (GLC) data are not directly comparable although they seem to be in the same range of recovery. The $^{14}$C data might reflect the parent compound plus any labeled metabolites; therefore, the recovery values may be higher than the GLC values.

A major metabolite was detected in the ethanol extract of 2,7-dichlorodioxin-treated soils. The metabolite was less mobile than DCDD in benzene–acetonitrile on TLC. The metabolite was eluted from the silica gel and methylated with diazomethane. The methylated metabolite was rechromatographed in benzene and migrated to the solvent front, suggesting a polar group on the non-methylated metabolite.

The soil persistence data suggested that TCDD is a relatively persistent compound. This is relative to various pesticides in the same concentration range. As pointed out by Kearney *et al.* (6), a concentration of 1 ppm of TCDD in soil is $10^6$ greater than the amount that would be found in soil receiving 2.24 kg/ha of 2,4,5-T with a contamination level of 1 ppm TCDD.

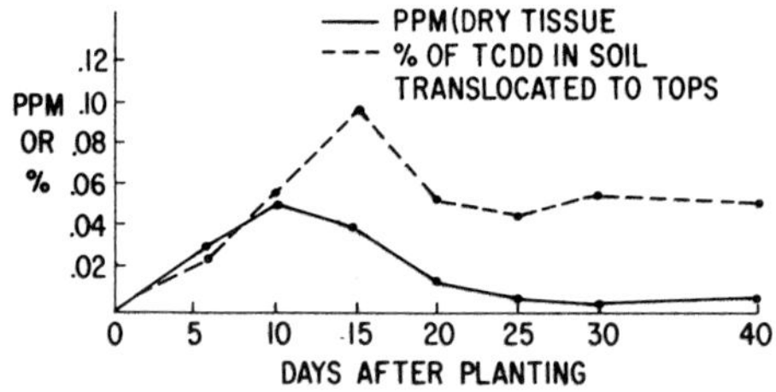


*Figure 3. Uptake of $^{14}$C-TCDD from Lakeland sandy loam by soybeans*

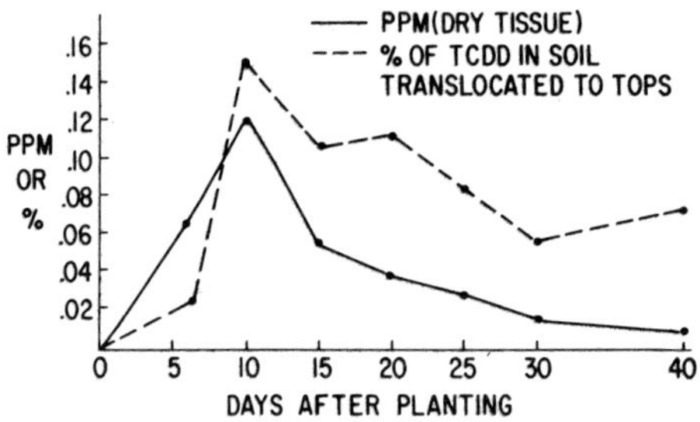


*Figure 4. Uptake of $^{14}$C-TCDD from Lakeland sandy loam by oats*

### *Plant Uptake*

Plant uptake is one of several routes by which an organic contaminant can enter man's food chain. The amount of uptake depends on plant species, concentration, depth of placement, soil type, temperature, moisture, and many other parameters. Translocation of the absorbed material into various plant parts will determine the degree of man's exposure—*i.e.*, whether the material moves to an edible portion of the plant. Past experience with nonpolar chlorinated pesticides suggested optimal uptake conditions are achieved when the chemical is placed in a soil with low adsorptive capacity (*e.g.*, a sand), evenly distributed throughout the soil profile, and with oil producing plants. Plant experiments were conducted with one set of parameters that would be optimal for uptake and translocation. The uptake of two dioxins and one phenol (2,4-dichlorophenol (DCP)) from one soil was measured in soybean and oats (7). The application rates were DCP = 0.07 ppm, DCDD = 0.10 ppm, and TCDD = 0.06 ppm. The specific activity of the compounds were: 5.8 $\mu C/mg$, 3.6 $\mu C/mg$, and 2.8 $\mu C/mg$, respectively. Oats and soybeans (at all growth stages) accumulated very small quantities of chlorodioxins or the chlorophenol (Figures 3 and 4). A maximum of 0.21% of the phenol and 0.15% of the dioxin present in soils was translocated to the aerial portion of oats and soybeans. No detectable amounts of $^{14}C$ were found in the grain or soybeans harvested at maturity. The amount of TCDD applied to these soils would be many thousands times greater than that which would occur in soils from herbicide applications containing a few parts per million of TCDD as an impurity. Even with these excessive rates in soil, significant amounts could not be measured in plants. Based on these studies, we must conclude that soil uptake of TCDD residues by plants is highly unlikely.

Next, we attempted to deal with translocation of foliar-applied TCDD. Labeled dioxins were applied to the center leaflet of the first trifoliate leaf of 3-week-old soybean plants and the first leaf blade of 12-day-old oat plants. All compounds were applied in an aqueous surfactant solution (Tween 80) to enhance leaf adsorption and to keep the water insoluble dioxins in solution. Plants were harvested 2, 7, 14, and 21 days after treatment, dissected into treated and untreated parts, and analyzed separately. Neither dioxin nor chlorophenol was translocated from the treated leaf. A rapid loss of the dichlorodioxin and dichlorophenol occurred from the leaf surface. This loss may have resulted from volatilization. Very little TCDD was lost from soybean leaves while a gradual loss (38% in 21 days) did occur from oat leaves.

### *Summary*

The detection of a potent dioxin impurity in a major herbicide has focused attention on the nature of chlorinated impurities in pesticides, and in a larger sense, impurities in all chlorinated industrial compounds used extensively in man's environment. The present 2,4,5-T controversy is overshadowed by the dioxin problem. Major disagreement still exists on their relative contributions to the teratogenic effects observed in chicks and the validity of interpretation of high dosage rates used to achieve these effects. We have avoided any assessment of the health-related aspects of dioxins but have dealt almost exclusively with dioxins as an environmental entity.

Information on dioxins in the environment was acquired rapidly by using some simple, but safe and reliable techniques developed for chlorinated pesticdes. Based on results of these tests, one should be able to predict whether routes of entry into aquatic and terrestrial food chains are significant, the rate and products of decomposition mechanism, and their general longevity in the environment.

Several facts have emerged from our studies with 2,7-DCDD and 2,3,7,8-TCDD. They are not biosynthesized by condensation of chlorophenols in soils, and they are not photoproducts of 2,4-dichlorophenol. They do not leach into the soil profile and consequently pose no threat to groundwater, and they are not taken up by plants from minute residues likely to occur in soils. Photodecomposition is insignificant on dry soil surfaces but is probably important in water. Dichlorodibenzo-*p*-dioxin is lost by volatilization, but TCDD is probably involatile. These compounds are not translocated within the plant from foliar application, and they are degraded in the soil.

## *Literature Cited*

1. Muelder, W. W., Shadoff, L. A., ADVAN. CHEM. SER. (1973) **120,** 1.
2. Woolson, E. A., Thomas, R. F., Ensor, P. D. J., *J. Agr. Food Chem.* (1972) **20,** 351.
3. Helling, C. S., Turner, B. C., *Science* (1968) **162,** 562.
4. Helling, C. S., *Soil Sci. Amer. Proc.* (1971) **35,** 737.
5. Plimmer, J. R., Klingebiel, U. I., *Science* (1971) **174,** 407.
6. Kearney, P. C., Woolson, E. A., Ellington, C. P. Jr., *Environ. Sci. Tech.* (1972) **6,** 1017.
7. Isensee, A. R., Jones, G. E., *J. Agr. Food Chem.* (1971) **19,** 1210.

RECEIVED February 8, 1972.

# 12

# Dioxin Residues in Lakeland Sand and Bald Eagle Samples

EDWIN A. WOOLSON and PETER D. J. ENSOR

Plant Science Research Division, Agricultural Research Service,
U. S. Department of Agriculture, Beltsville, Md. 20705

WILLIAM L. REICHEL

U. S. Department of Interior, Patuxent Wildlife Center, Laurel, Md. 20810

ALVIN L. YOUNG[1]

U. S. Air Force, Eglin Air Force Base, Fla. 32542

*2,3,7,8-Tetrachlorodibenzo-*p*-dioxin (TCDD) is a possible contaminant of 2,4,5-trichlorophenoxyacetic acid (2,4,5-T). Higher chlorinated dioxins are possible contaminants of pentachlorophenol. To assess the ecological importance of chlorinated dioxins, eagle tissue was examined by electron-capture gas chromatography for the presence of dioxins. No dioxins were detected at a minimum detection limit of 50 ppb. Eagle samples from various regions in the United States were included. Soil samples from a Lakeland sand in Florida were analyzed for TCDD after applying up to 947 lbs 2,4,5-T/acre (between 1962–1969). No TCDD was detected at a minimum limit of <1.0 ppb in core samples to a 3 foot depth. Small amounts (<20 ppb) of 2,4-D and 2,3,5-T were found at all depths.*

2,3,7,8-Tetrachlorodibenzo-*p*-dioxin (TCDD) is a contaminant in 2,4,5-trichlorophenoxyacetic acid (2,4,5-T) which may occur in the manufacturing process. Before 1971 some samples of 2,4,5-T contained from 2–50 ppm TCDD (*1*) in the technical acid. Highest concentrations of TCDD occurred before 1968.

At Eglin AFB, Florida, the U.S. Air Force initiated an experiment in 1962 to test aerial application equipment for deposition of defoliants

[1] Present address: U.S. Air Force Academy, Colo. 80840.

**Table I. Approximate Deposition Rate of Herbicides (pounds active ingredient/acre) Applied to Test Area C-52A, Eglin AFB Reservation, Florida**

| *Test Grid* | *Grid Area*[a] *(Acres)* | *Herbicide (Pounds Active Ingredient/Acre)* 2,4-D | 2,4,5-T |
|---|---|---|---|
| Grid 1 | 92 | 947 (1962-64)[b] | 947 (1962-64) |
| Grid 2 | 92 | 584 (1964-66) | 584 (1964-66) |
| Grid 3 | 240 | 183 (1968-69) | 160 (1968-69) |
| Grid 4 | 92 | 30 (1967) | |

[a] Grids 2 and 4 fall within the confines of the 640-acre Grid 3, however, the positioning of the test arrays on Grid 3 has resulted in most of the agent being disseminated within a 240 acre area with only slight infringement on the original sites of Grids 2 and 4.
[b] Years when the majority of the herbicide was applied.

containing 2,4-D, 2,4,5-T, dimethyl arsinic acid (cacodylic acid), and picloram. These chemicals were sprayed over 92 and 240 acre plot areas during an 8-year period. Some plots were sprayed through 1970. The approximate application rates and years of application are shown in Table I. The heaviest rates of 2,4,5-T (947 lbs/acre) were applied during 1962–64 while the lightest rate was in 1968–69.

Tissue extracts from 19 bald eagle (*Haliaectas leucocephalus*) carcasses were examined to determine if biomagnification of TCDD had occurred in a manner similar to DDT. These carcasses came from the states of Alaska, Maine, North Dakota, Wisconsin, Michigan, Minnesota, Arkansas, Illinois, Missouri, Maryland, Virginia, Iowa, New York, New Jersey, and Florida between 1966 and 1971 and were collected and furnished by scientists at the Patuxent Wildlife Center, U.S. Department of the Interior, Laurel, Md. The samples were selected from these states to provide a widely dispersed sampling population.

This study was done to determine if TCDD residues could be detected in soil receiving massive doses of 2,4-D and 2,4,5-T or in bald eagle extracts, as a representative of the top of a food chain.

## *Materials and Methods*

Twelve, 36-inch soil cores of the Lakeland sand were selected for chemical analyses in December 1970. Twenty-five gram samples of 6-inch increments were acidified and extracted with 1:1 hexane:acetone. Each sample was extracted with 1*N* KOH, and the aqueous phase was saved for 2,4-D and 2,4,5-T analysis. The hexane phase was extracted repeatedly with concentrated $H_2SO_4$ until the acid was clear. The $H_2SO_4$ was removed, and the extract was drained through $NaHCO_3$ and anhydrous

$Na_2SO_4$ after a water wash. The volume was adjusted, and an aliquot equivalent to 50 or 100 mg of soil was injected on to a 5% OV-225 GC column (*1*). Samples were amended at 1 and 5 ppb TCDD for recovery studies.

The alkaline solution containing 2,4-D and 2,4,5-T residues was acidified and extracted with diethyl ether. $NH_4OH$ (0.1 ml) was added to the extract which was then taken to dryness. The residue was butylated with 14% $BCl_3$ in *n*-BuOH following the concentrated-$BF_3$ method described by Woolson and Harris (*2*).

Eagle tissue was prepared and extracted by the methods described by Mulhearn *et al.* (*3*). The extract was treated with $H_2SO_4$, passed through a 5 cm × 19 mm alumina column (Fisher A-540 or equivalent) with diethyl and petroleum ether, and treated with 1:1 $HNO_3$:$H_2SO_4$ at 0°C after drying. The acids were added to ice water, and hexane was added. The hexane layer was drained through a $NaHCO_3$–$Na_2SO_4$ mixed column. The volume was adjusted, and the sample was tested for TCDD and the higher polychlorinated dioxins. Samples were amended at 50 ppb TCDD and hexa-dioxin for recovery studies.

### *Results and Discussion*

The soil analysis is presented in Table II. Small amounts of 2,4-D and 2,4,5-T were detected in soil samples receiving these herbicides. Background values from the control soils were subtracted from the observed values in treated soils. The samples were not corrected for recovery since it was better than 80% for the method. Residues decreased with time after application. Leaching and microbial decomposition could account for this observation.

No TCDD could be detected in soil with a detection limit of less than 1 ppb. Gas chromatographic (GC) traces at the left of Figure 1 represent a soil which would have contained 5 ppb of TCDD; traces at the right are the unamended soil extracts. Arrows indicate the position where TCDD would occur. At the bottom right is a trace of the extract from a soil amended at the 1 ppb TCDD level.

Treatment of the extract with UV light (*4*) for 16 hours completely destroyed TCDD in the amended soil. The extract from the unamended soil was not changed by irradiation. Peaks close to the TCDD peak in retention time (±0.2 min) were not altered. Recovery of *ca.* 100% was obtained when 5 ppb TCDD was added to the soil. A peak should be discernible at a concentration of 1 ppb as seen on the control core. However, background interference, even in the cleaned up residue, increased tremendously at the very low levels, and confirmation of a peak's identity was very difficult.

UV irradiation was used as a confirmation technique on any peak suspected of being TCDD (±0.2 min); no TCDD could be detected at less than 1 ppb. The question then arises, what level of contamination had to be present in the original 2,4,5-T to detect 1 ppb TCDD in soil.

**Table II. Analysis of Lakeland Sand for 2,4-D, 2,4,5-T, and TCDD**

| *Application Rate and Date* | *Core Depth* | *2,4-D, ppb* | *2,4,5-T, ppb* | *TCDD, ppb* |
|---|---|---|---|---|
| 947 lb/Acre (1962-64) | 0-6 [a] | 0.4 | 0.8 | < 0.1 |
| | 6-12 | < 0.1 | 1.2 | < 0.1 |
| | 12-18 | < 0.1 | 0.8 | < 0.2 |
| | 18-24 | 0.4 | 0.3 | < 0.2 |
| | 24-30 | < 0.1 | 4.4 | < 0.4 |
| | 30-36 | < 0.1 | 0.5 | < 0.2 |
| Total (% of applied) | | 0.8 (0.0002) | 8.0 (0.002) | |
| 584 (lb/Acre (1964-66) | 0-6 [b] | 0.9 | 3.0 | < 0.1 |
| | 6-12 | 0.3 | 0.6 | < 0.2 |
| | 12-18 | < 0.1 | 0.6 | < 0.2 |
| | 18-24 | 0.3 | 1.1 | < 0.2 |
| | 24-30 | 0.8 | 1.4 | < 0.2 |
| | 30-36 | < 0.1 | < 0.1 | < 0.4 |
| Total (% of applied) | | 2.3 (0.0008) | 6.7 (0.0023) | |
| 160 lb/Acre—2,4,5-T plus 183 lb/Acre—2,4-D (1968-69) | 0-6 [c] | 3.2 | 3.4 | < 0.2 |
| | 6-12 | 3.7 | 0.8 | < 0.2 |
| | 12-18 | 1.4 | 0.8 | < 0.2 |
| | 18-24 | 3.3 | 1.2 | < 0.2 |
| | 24-30 | 0.9 | 0.7 | < 0.3 |
| | 30-36 | 2.9 | 1.5 | < 0.3 |
| Total (% of applied) | | 15.4 (0.02) | 8.4 (0.01) | |

[a] Average of 2 cores.
[b] Average of 4 cores.
[c] Average of 6 cores.

If we assume that the TCDD is contained in the surface 6 inches of the soil profile since it is relatively immobile (*5*), then the 2,4,5-T at the 947 lbs of active ingredient/acre treatment would have had to contain 2.1 ppm TCDD to be observed. At the lower application rates of 584 and 160 lbs/acre, the 2,4,5-T would have had to contain 3.5 and 12.5 ppm TCDD in the technical materials to have 1 ppb in the top 6 inches of soil. Since the soil is sandy and high rainfall occurred in the area, maximum movement of materials in soil may occur causing TCDD to be present deeper in the profile. If the TCDD moved uniformly throughout the 36 inch soil profile, then six times more TCDD would have had to be present in the original 2,4,5-T for detection. This would have required the presence of 12.6, 21.0, and 75.0 ppm TCDD in the 2,4,5-T applied in the three treatments. These calculations are based on the assumption that no degradation occurred in or on the soil.

A survey of TCDD content in 2,4,5-T samples (*1*) revealed that levels of 30–40 ppm TCDD were not uncommon before 1969. Therefore, it is not unreasonable to expect residue in the soil environment within the surface 36 inches of soil at the two higher treatment rates. The fact

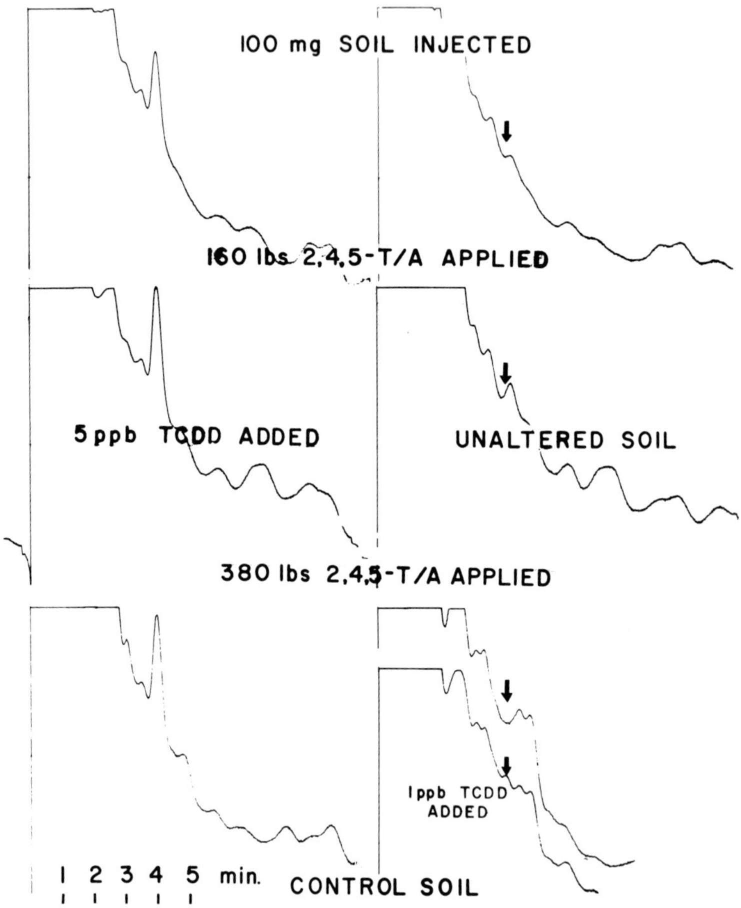


*Figure 1. Extracts from three TCDD-amended and non-amended Lakeland Sand soils treated with 2,4-D and 2,4,5-T*

*Column conditions: length, 1.8-m × 4-mm glass; solid support, Chromosorb W, 80–100 mesh; injector, 240°C; column, 200°C; detector, $Ni^{63}$, 310°C; flow rate, 80–100 ml/min*

that we did not find TCDD in any core at any depth indicates one of several things:

(1) The 2,4,5-T applied contained less than 2 ppm TCDD.

(2) The TCDD is moved deeper than 36 inches in the soil.

(3) The TCDD is decomposed on or in the soil either photochemically or biologically.

(4) Wind erosion has removed the TCDD from the point of application.

Any or all of the conditions above could contribute to the lack of TCDD residues in this soil at a level greater than 1 ppb.

Bald eagle extracts were examined primarily for TCDD. However, if hexa-, hepta-, and octadioxins were present, they could have been detected. The GC trace of the cleaned-up extract (Figure 2) is presented along with a trace of the extract amended with 0.05 ppm of TCDD and 0.05 ppm of a hexadioxin isomer. The cleanup method was the same as that for the soils presented earlier with the addition of an alumina column for partitioning the extract with petroleum and diethyl ethers. There was a slight peak in the valley on the amended sample and no peak in the original extract at the TCDD position. The second arrow to the right indicates the position of the hexadioxin. The original trace has a peak with a slightly shorter retention time than the hexa-dioxin. The large peaks further out are reasonably close to the two hepta- and the octadioxin peaks. To confirm possible identity, we subjected the extract to UV irradiation for 16 hours (trace 3).

The amended TCDD peak completely disappeared; the hexa-peak remained, and the possible hepta- and octadioxin peaks disappeared. Since the authentic compounds are destroyed only slightly by irradiation (*4*), this absence of the peaks indicates that the peaks near the hepta- and octadioxins were not dioxins. The peaks could have represented

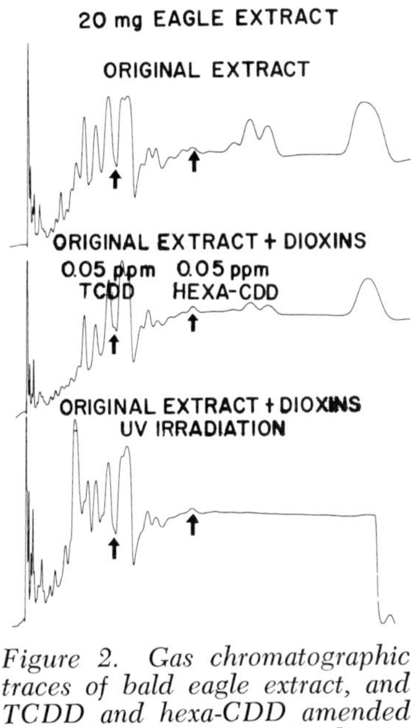


*Figure 2. Gas chromatographic traces of bald eagle extract, and TCDD and hexa-CDD amended extract before and after irradiation by UV light*

*Conditions:* see *Figure 1; column, 230°C for 4 min, 230°–260°C at 4 C/min, hold at 260°C*

chlorinated furans, PCB's, or some other related compounds, which are destroyed by UV light. Identification of these peaks was not attempted.

No dioxin residues were detected at a level of 0.05 ppm TCDD, the lower limit of detection for most pesticides in tissue samples run by the Patuxent Wildlife Research Center. The non-detection of dioxin residues can imply several things:

(1) There is no dioxin build-up in the food chain.

(2) The build-up is less than the current detectable level of 50 ppb.

(3) The eagles examined were not contaminated although other samples might be.

(4) Other species could feed on a different food chain to accumulate dioxins.

### *Acknowledgment*

The authors would like to thank Dow Chemical Co. for the $^{14}C$-dioxins used in methodology development and David Firestone, Food and Drug Administration, for supplying the analytical standards.

### *Literature Cited*

1. Woolson, E. A., Thomas, R. F., Ensor, P. D. J., *J. Agr. Food Chem.* (1972) **20**, 351.
2. Woolson, E. A., Harris, C. I., *Weeds* (1967) **15**, 168.
3. Mulhern, B. M., Reichel, W. L., Locke, L. N., Lamont, T. G., Belisle, A., Cromartie, E., Bagley, G. E., Prouty, R. M., *Pest. Monit. J.* (1970) **4**, 141.
4. Crosby, D. G., Wong, A. S., Plimmer, J. R., Woolson, E. A., *Science* (1971) **173**, 748.
5. Helling, C. S., *Soil Sci. Soc. Amer. Proc.* (1971) **35**, 737.

RECEIVED February 18, 1972.

13

# The Stability of Pentachlorophenol and Chlorinated Dioxins to Sunlight, Heat, and Combustion

R. H. STEHL, R. R. PAPENFUSS, R. A. BREDEWEG, and R. W. ROBERTS

Analytical Laboratories, The Dow Chemical Co., Midland, Mich. 48640

*Pentachlorophenol in field-use conditions is subjected to environmental factors leading to its degradation, resulting theoretically in chlorinated dibenzo-*p*-dioxin formation. Analysis of combustion products of wood and paper treated with pentachlorophenol indicated no increase and possibly a decrease in octachlorodibenzo-*p*-dioxin concentration while paper treated with sodium pentachlorophenate increased slightly in octachlorodibenzo-*p*-dioxin concentration after combustion. Photolysis of sodium pentachlorophenate under different sources of artificial sunlight and natural daylight resulted in severe degradation of the pentachlorophenate with only trace amounts of octachlorodibenzo-*p*-dioxin formed. 2,7-Dichloro- and 2,3,7,8-tetrachlorodibenzo-*p*-dioxins are subject to rapid photolytic decomposition under artificial sunlight while octachlorodibenzo-*p*-dioxin shows considerably more stability. 2,3,7,8-Tetrachlorodibenzo-*p*-dioxin is somewhat stable to 700°C while decomposition is complete at 800°C.*

Over the past several years concern over the contamination of our environment has increased. This interest has caused industry to examine more closely the effects its products might have on our surroundings. Not only are the possible effects of the products themselves being considered, but it is becoming increasingly necessary to include the effects of impurities contained in these materials, as well as the effects of the degradation products formed by environmental forces such as heat, fire, and sunlight.

This paper discusses the effects of the pyrolysis of wood and paper treated with pentachlorophenol or sodium pentachlorophenate along with the exposure of sodium pentachlorophenate to ultraviolet light. The effect of heat, light, and air with respect to some of the chlorinated dibenzo-*p*-dioxins will be considered.

## *Methods*

**Pyrolysis of Wood or Paper Treated with Pentachlorophenol or Sodium Pentachlorophenate.** The treated sample was combusted in a modified 250-ml Erlenmeyer flask that had been fitted with a glass grate. The volatile combustion products were captured in cold traps immersed in an ice bath, dry ice, and liquid nitrogen (connected in series). Combustion residues, including ashes, were dissolved in benzene. The benzene solution containing the combustion products was washed with 1*N* sodium hydroxide solution, sulfuric acid, and sulfuric acid containing a small amount of water. Finally, the washed solution was passed through a silica gel column using 20% benzene–hexane (v/v) solution as the eluant. Octachlorodibenzo-*p*-dioxin in the fraction was determined by gas chromatography using a Varian 1200 gas chromatograph fitted with a 7-foot 1/8 inch 5% QF1 on 60/80 mesh Gas Chrom Q column equipped with either electron capture or flame ionization detectors.

**Heat Stability of 2,3,7,8-Tetrachlorodibenzo-*p*-dioxin.** A sample boat containing 1–3 mg of 2,3,7,8-tetrachlorodibenzo-*p*-dioxin was placed in a Sargent Tube Heater (Model J-807) which contained a 1/2 inch i.d. quartz tube. Air was aspirated through the tube at a velocity of 0.02 ft/sec. This velocity gave a residence time of 21 sec. (Residence time was extended by connecting a second heater in series with the first.)

Decomposition products were captured in two traps immersed in dry ice followed by a third trap containing 5 ml of benzene. After 1 hour, the boat, tube, and traps were rinsed with benzene. The benzene solution was analyzed by gas chromatography using an electron capture detector to determine the concentration of unreacted 2,3,7,8-tetrachlorodibenzo-*p*-dioxin.

**Light Stability of Sodium Pentachlorophenate.** Pentachlorophenol was dissolved in a high capacity pH 8 phosphate buffer such that the final concentration of sodium pentachlorophenate was 500 ppm. The pentachlorophenate solutions were placed in quartz vessels fitted with ground, quartz stoppers and exposed to various sources of ultraviolet light as well as daylight. Two 20-watt fluorescent black lights were placed 2 inches from the liquid's surface. A General Electric R.S. sunlamp and a high intensity mineral light emitting at 254 nm were placed 11 inches from the surface of their respective solutions. The flask exposed to daylight was suspended 20 inches above the laboratory roof. Unreacted sodium pentachlorophenate was determined by ultraviolet spectroscopy. Octachlorodibenzo-*p*-dioxin was separated by extraction with three successive 50-ml portions of benzene. The combined benzene layers were washed with 0.1*N* sodium hydroxide solution, 0.1*N* hydrochloric acid solution, and deionized water. Octachlorodibenzo-*p*-dioxin was determined by gas chromatography using an electron capture detector.

**Light Stability of Chlorinated Dibenzo-*p*-dioxins.** A known concentration of chlorinated dibenzo-*p*-dioxin was placed in a quartz 1-cm path length cell. A General Electric R.S. sunlamp was placed either 0.5 or 1.0 meters from the face of the cell. The stability of the chlorinated dibenzo-*p*-dioxin was monitored using ultraviolet spectroscopy.

**Stability of 2,3,7,8-Tetrachlorodibenzo-*p*-dioxin Towards Air Oxidation Under Simulated Conditions.** Air was bubbled through two borosilicate glass gas absorption bottles equipped with fritted glass bubblers. The first bottle contained 1-octanol for presaturation of the air, and the second bottle contained 1-octanol solutions of the dioxin treated as follows: (1) octanol only, (2) octanol mixed with 74–105$\mu$ glass beads to increase the surface area, and (3) octanol mixed with magnesium oxide to simulate a basic soil. The original solution and the sample solutions were scanned with a UV spectrophotometer at various time intervals for 4 days to determine the stability of 2,3,7,8-tetrachlorodibenzo-*p*-dioxin.

**Table I. Octachlorodibenzo-*p*-dioxin Concentration before and after Burning Paper and Wood Treated with Pentachlorophenol and Sodium Pentachlorophenate**

| *Sample* | *Pentachlorophenol in Unburned Sample, %* | *Octachlorodibenzodioxin, ppm* | |
|---|---|---|---|
| | | *In Sample* | *In Combustion Products*[a] |
| Wood | 0.048, 0.050 | 0.04, <0.02 | 0.04, <0.02 |
| Wood + Penta | 1.0, 1.0 | 43, 48 | 1.7, 0.95 |
| Wood + Pure Penta[b] | 6.0, 6.0 | 0.04, 0.03 | 0.10, <0.01 |
| Paper | <0.01 | <0.02 | <0.01 |
| Paper + Penta | 0.19 | 75 | 24 |
| Paper + Pure Penta | 0.76 | <0.02 | <0.01 |
| Paper + Na Penta | 2.7 | 13 | 69 |
| Paper + Pure Na Penta[c] | 3.4 | <0.02 | 45 |

[a] Combustion product weight was calculated assuming the samples contained 30% water.
[b] Purified pentachlorophenol.
[c] Purified sodium pentachlorophenate.

## *Discussion and Results*

**Effects of Combustion and Heat.** The results of combusting wood and paper treated with pentachlorophenol or sodium pentachlorophenate are shown in Table I. These results indicate that octachlorodibenzo-*p*-dioxin concentration did not increase as the result of combusting either wood or paper treated with pentachlorophenol. It seems that the concentration of octachlorodibenzo-*p*-dioxin concentration was actually decreased during combustion. However, paper treated with sodium pentachlorophenate did increase in octachlorodibenzo-*p*-dioxin concentration as the result of combustion.

**Table II. Decomposition of 2,3,7,8-Tetrachlorodibenzo-*p*-dioxin after Exposure to Elevated Temperatures**

| *Temperature, °C* | *% Decomposition (21 Sec Exposure)* | *% Decomposition (50 Sec Exposure)* |
|---|---|---|
| 500 | 39 | 42 |
| 600 | 40 | 59 |
| 700 | 50 | 53 |
| 800 | 99.5 | |

The heat stability of 2,3,7,8-tetrachlorodibenzo-*p*-dioxin is shown in Table II. The material is quite stable to 700°C with 50% decomposition obtained at this temperature. More than doubling the residence time resulted in only slightly more decomposition. However, decomposition of 2,3,7,8-tetrachlorodibenzo-*p*-dioxin is complete at 800°C.

**Effect of UV Radiation on Sodium Pentachlorophenate.** Sodium pentachlorophenate is quickly degraded in the presence of ultraviolet light as well as daylight, as shown in Table III. The degradation is almost complete after 16 hours exposure to the General Electric R.S. sunlamp. Three and one-half days of rainy and overcast skies account for the somewhat lower degradation observed in the sample exposed to daylight.

**Table III. Stability of Sodium Pentachlorophenate to Sunlight and Artificial Sunlight in pH 8 Phosphate Buffer**

| *Light Source* | *Exposure Time* | *% Degradation* |
|---|---|---|
| Control | 0 | |
| Fluorescent black light | 16 hrs. | 7.8 |
| Mineral light | 16 hrs. | 18.7 |
| General Electric RS sunlamp | 16 hrs. | 93.9 |
| Daylight | 5 days | 45.6 |

**Table IV. Octachlorodibenzo-*p*-dioxin in Irradiated Solutions in pH 8 Phosphate Buffer**

| *Light Source* | *Exposure Time* | *Octachlorodibenzo-p-dioxin Formed Based on Total Sample Weight* |
|---|---|---|
| Control | 0 | 0.005%[a] |
| Fluorescent black light | 16 | 0.024% |
| Mineral light | 16 | 0.024% |
| General Electric RS sunlamp | 16 | 0.034% |
| Daylight | 5 days | 0.029% |

[a] Initially present

A small amount of octachlorodibenzo-*p*-dioxin, amounting to approximately 0.03%, was formed during the photolysis of sodium pentachlorophenol buffered at pH 8 as shown by the data in Table IV.

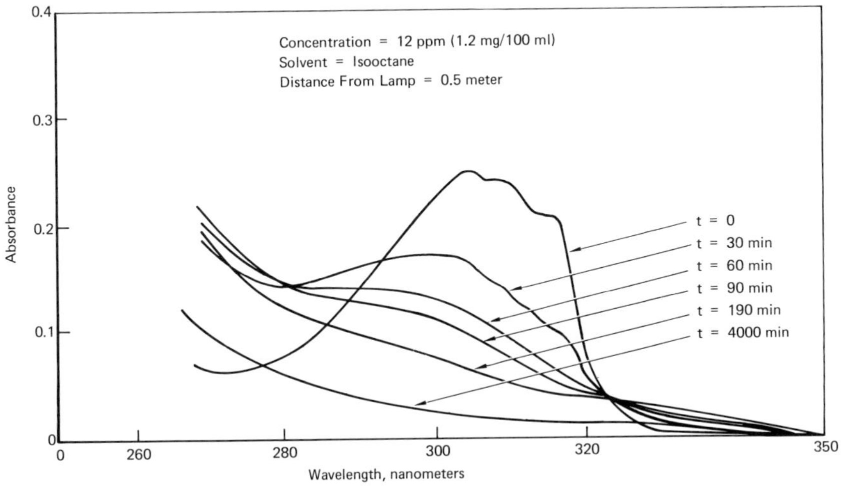


*Figure 1. Effect of sunlight on 2,3,7,8-tetrachlorodibenzo-*p*-dioxin*

**Effect of UV Irradiation and Air Oxidation on Chlorinated Dibenzo-*p*-dioxin.** The spectrum of 2,3,7,8-tetrachlorodibenzo-*p*-dioxin in isooctane is shown in Figure 1 along with rapid decrease in the UV spectrum of 2,3,7,8-isomer upon exposure to UV light. The regular decrease in the spectrum observed in Figure 1 suggested a first order decay for the 2,3,7,8-tetrachlorodibenzo-*p*-dioxin. The negative log of the absorbance was therefore plotted as a function of time as shown in Figure 2.

The half-life of the 2,3,7,8-tetrachlorodibenzo-*p*-dioxin in isooctane was estimated to be 40 min for the 0.5 meter exposure and 3 hours for the one meter exposure. The half-life in 1-octanol was essentially the same. The 24-hour photolysis products of the 2,3,7,8-tetrachlorodibenzo-*p*-dioxin were examined by gas chromatography. The smallest concentration of 2,3,7,8-tetrachlorodibenzo-*p*-dioxin that could be detected by the instrument was 0.5 ppm. When an injection of the 24-hour photolysis product was made, no tetra was detected. An additional confirmation of the disappearance of the 2,3,7,8-tetrachlorodibenzo-*p*-dioxin in the 24-hour photolysis products was obtained when the material was submitted to the Chemical-Biology Research Laboratory for rabbit testing. No chloracnegenic activity was indicated.

Under similar conditions a half-life of 40 min was determined for 2,7-dichlorodibenzo-*p*-dioxin in both solvents.

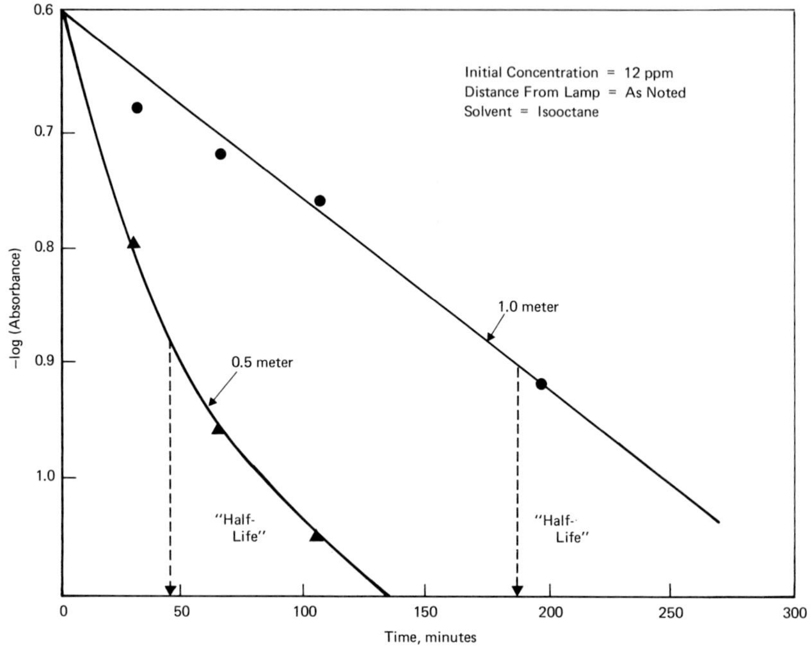


*Figure 2. Rate of disappearance of 2,3,7,8-tetrachlorodibenzo-*p*-dioxin in artificial sunlight*

Very little change was observed in the octachlorodibenzo-*p*-dioxin on exposure to artificial sunlight. Over extended periods of time (18–24 hours) there was some evidence of decay. Approximately 20% photolysis was observed in isooctane at the end of 18 hours and about 6% photolysis of the octachlorodibenzo-*p*-dioxin after 20 hours exposure in 1-octanol.

Periodic examination of the UV spectra of 2,3,7,8-tetrachlorodibenzo-*p*-dioxin under simulated air oxidation conditions indicated no change in the UV spectra. Therefore, the 2,3,7,8-isomer is probably stable toward air oxidation.

## *Conclusion*

Combustion of wood or paper treated with pentachlorophenol resulted in no increase and more probably a decrease in octachlorodibenzo-*p*-dioxin concentrations while octachlorodibenzo-*p*-dioxin increased slightly in paper treated with sodium pentachlorophenate. The photolytic degradation of sodium pentachlorophenate at pH 8 is very rapid. Under these controlled conditions formation of no more than 0.03% octachlorodibenzo-*p*-dioxin was observed. The 2,3,7,8 isomer, one of the most active chloracnegens is seemingly stable towards air oxidation but was destroyed at 800°C and was, along with the 2,7 isomer, unstable in the presence of UV light with the 2,7- and 2,3,7,8-chlorinated dibenzo-*p*-dioxins having a half-life of 40 and 45 min, respectively. The octachlorodibenzo-*p*-dioxin seems stable towards UV radiation. However, tests have shown that this material is relatively innocuous.

RECEIVED February 8, 1972.

# 14

# Preparation of Chlorodibenzo-p-dioxins for Toxicological Evaluation

ORM ANILINE

Organic Chemicals Production Research Laboratory, The Dow Chemical Co., Midland, Mich. 48640

*Chlorinated dibenzo-*p*-dioxins were prepared on the gram scale for use as toxicological standards. 2,7-Dichlorodibenzo-*p*-dioxin was prepared by catalytic condensation of potassium 2-bromo-4-chlorophenate in 70% yield. Thermal condensation of the potassium salt of 2,4,4′-trichloro-2′-hydroxydiphenyl ether gave a mixture of the 2,8- and 2,7-dichlorodibenzo-*p*-dioxins which were separated by fractional recrystallization. 2,3,7,8-Tetrachlorodibenzo-*p*-dioxin of 99.9+% purity was prepared by catalytic condensation of potassium 2,4,5-trichlorophenate. An isomeric mixture of hexachlorodibenzo-*p*-dioxins was prepared by pyrolytic condensation of sodium 2,3,4,6-tetrachlorophenate. Chlorination of pentachlorophenol (containing <0.07% tetrachlorophenol) in trichlorobenzene gave octachlorodibenzo-*p*-dioxin in 80% yield contaminated by 5–15% heptachlorodibenzo-*p*-dioxin. Oxidative methods were used to produce octachlorodibenzo-*p*-dioxin at 99.9% purity.*

Interest in the toxicology of chlorinated dibenzo-*p*-dioxins has created a need for gram quantities of pure material. Although chlorodibenzo-*p*-dioxins have been prepared for analytical characterization (*1*) and for qualitative determination of biological activity, suitable toxicological standards have required improved synthetic methods and purification techniques. For example, the synthetic approaches based on chlorination of the parent hydrocarbon, or of partially chlorinated dibenzo-*p*-dioxins, are not selective enough to produce uncontaminated product (*2, 3*). Separation techniques for isolating desired chlorodibenzo-*p*-dioxin from mixtures containing other chlorodibenzo-*p*-dioxins generally cannot be used to prepare pure material on the gram scale. This report describes the preparation of several chlorodibenzo-*p*-dioxins of 99+% purity.

### *Discussion*

Preparations of 2,7-dichlorodibenzo-*p*-dioxin based on the condensation of sodium 2,4-dichlorophenate have typical yields of less than 20% (*4, 5, 6, 7*). The formation of polymeric side-reaction products under these conditions suggests preferential nucleophilic attack at the para position (Reaction 1).

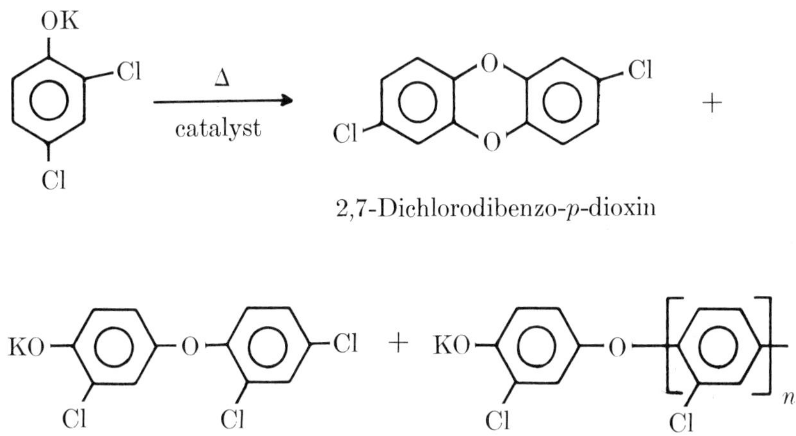


(1)

In contrast to this low yield reaction, condensation of potassium 2-bromo-4-chlorophenate under Ullmann conditions gave 2,7-dichlorodibenzo-*p*-dioxin in 74% yield (*8*). The preferential displacement of bromine follows the pattern Weingarten observed in the reaction between potassium phenate and bromobenzene (*9, 10*).

The synthetic preparation of 2,8-dichlorodibenzo-*p*-dioxin was facilitated in that the chemical precursor, 2,4,4′-trichloro-2′-hydroxydiphenyl ether, was available as a pure material. Condensation was induced by heating the potassium salt at 200°C for 15 hours in bis(2-ethoxyethyl) ether. Product analysis by GLC and mass spectrometry revealed an unexpected dichlorophenol and a monochlorodibenzo-*p*-dioxin. Further, the product initially isolated by crystallization from the reaction mixture was 2,7-dichlorodibenzo-*p*-dioxin, rather than the expected 2,8-isomer. Cooling of the mother liquor yielded crystalline plates which were shown to be 2,8-dichlorodibenzo-*p*-dioxin by x-ray diffraction (Reaction 2).

Under the GLC techniques used, the 2,8- and 2,7-dichlorodibenzo-*p*-dioxins were not resolved, so their relative abundance was not determined. Although our data do not explain the ease with which the

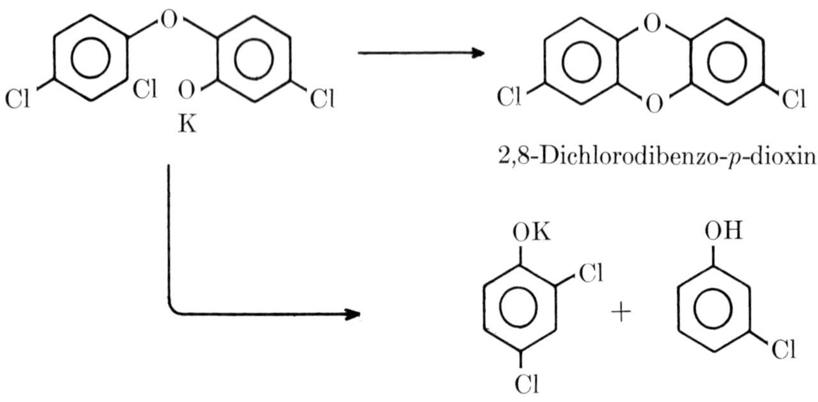

2,8-Dichlorodibenzo-*p*-dioxin

(2)

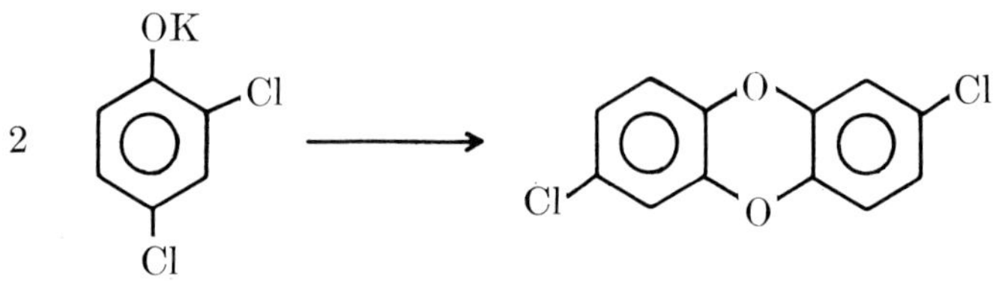

2,7-Dichlorodibenzo-*p*-dioxin

2,7-dichlorodibenzo-*p*-dioxin formed under non-catalytic conditions, it does give evidence that the formation of hydroxydiphenyl ethers from 2-halophenols may be reversible. Formation of hydroxydiphenyl ethers from dibenzo-*p*-dioxins under these conditions has not been experimentally demonstrated.

The preparation of 2,3,7,8-tetrachlorodibenzo-*p*-dioxin by chlorination of 2,7-dichlorodibenzo-*p*-dioxin yields a product containing significant quantities of trichloro- and pentachlorodibenzo-*p*-dioxins (*11*). Such mixtures are not amenable to separation on a preparative scale. Although 2,3,7,8-tetrachlorodibenzo-*p*-dioxin has been prepared by the pyrolytic condensation of sodium 2,4,5-trichlorophenate, this method is undesirable for preparation of 2,3,7,8-tetrachlorodibenzo-*p*-dioxin on the gram scale (*2, 12*). The pyrolytic reaction is difficult to control and the potential danger is enhanced by the product's toxicity. The salt was dissolved in bEEE [bis(2-ethoxyethyl) ether, bp 189°–190°C] and refluxed for 15 hours with the Ullmann catalyst. The desired product was obtained in 39% yield by condensation of potassium 2,4,5-trichlorophenate (Reaction 3).

To investigate the possibility of preparing uniquely substituted hexachlorodibenzo-*p*-dioxins, several experiments were performed under Ullmann conditions. Although condensation of potassium 2-bromo-4-chlorphenate occurred readily, several 2,6-substituted phenates failed to

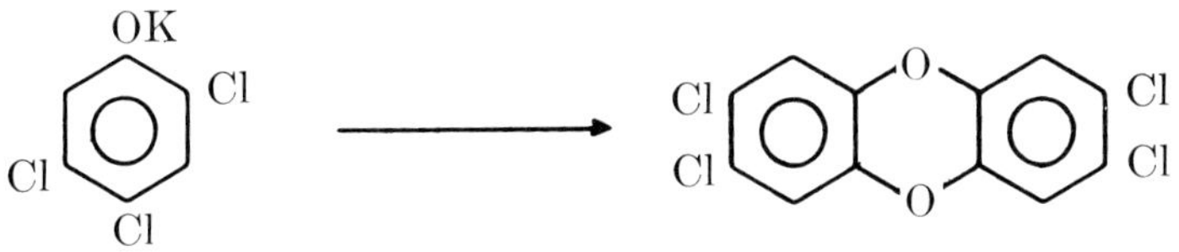


(3)

2,3,7,8-Tetrachlorodibenzo-*p*-dioxin

yield the desired product. The reactions attempted are summarized in Table I. In each case the appearance of the reaction mixture indicated polymer formation which was not observed during the preparation of 2,7-dichlorodibenzo-*p*-dioxin and 2,3,7,8-tetrachlorodibenzo-*p*-dioxin. Tomita's qualitative test for hydrogen-bearing dibenzo-*p*-dioxins gave a negative result for samples drawn from the reaction mixture (*2*). These preliminary reactions were not analyzed by the GLC techniques used for the other products in this report. Preparation of 1.5 grams of an isomeric mixture of 99+% pure hexachlorodibenzo-*p*-dioxins was accomplished by pyrolytic condensation of sodium 2,3,4,6-tetrachlorophenate. The product contained two isomeric hexachlorodibenzo-*p*-dioxins, deter-

**Table I. Attempts to Prepare Hexachlorodibenzo-*p*-dioxins Under Catalytic Conditions**[a]

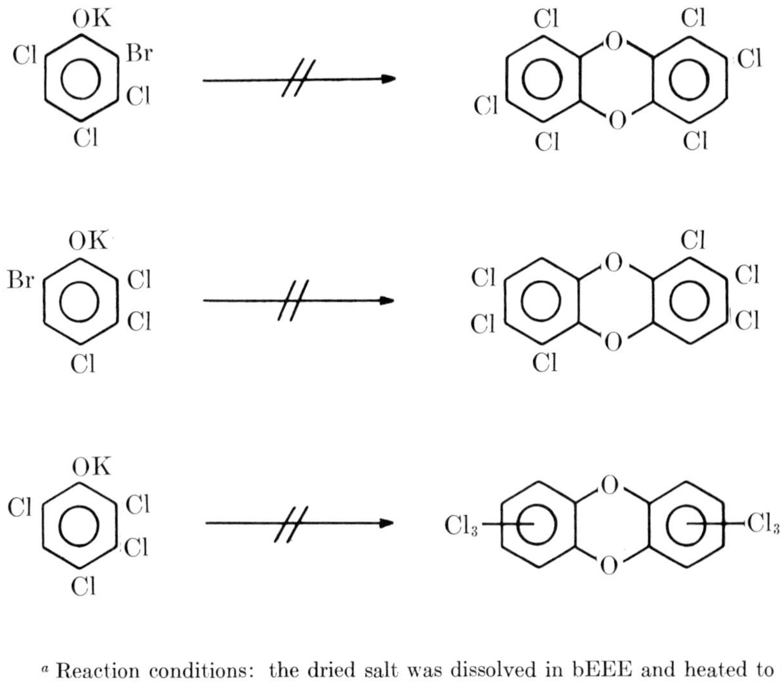

[a] Reaction conditions: the dried salt was dissolved in bEEE and heated to reflux for 15 hours after addition of the copper catalyst.

mined by GLC. Isolation of individual isomers and structural determination by x-ray crystallography was not attempted.

Extreme caution should be exercised in preparing hexachlorodibenzo-*p*-dioxins under the conditions described in the Experimental section; the product seems to form during controlled exotherms at 325°C. The suspected toxicity of this material and possible experimental flaws could place the experimenter in danger.

The most convenient and successful synthetic preparation of octachlorodibenzo-*p*-dioxin has been described by Kulka (*13*). The procedure involves chlorination of pentachlorophenol in refluxing trichlorobenzene to give octachlorodibenzo-*p*-dioxin in 80% yield. Kulka has explained the reaction as coupling between two pentachlorophenoxy radicals. Large amounts (5–15%) of heptachlorodibenzo-*p*-dioxin were observed in the unpurified product. Since the pentachlorophenol used in this study contained 0.07% tetrachlorophenol, we feel that tetrachlorophenol may be produced *in situ* (Reaction 4). Such a scheme would be analogous to the formation of 2,4-dichlorophenol and 3-chlorophenol produced from 2,4,4′-trichloro-2′-hydroxydiphenyl ether (Reaction 2). The solubility of octachlorodibenzo-*p*-dioxin was determined in various solvents; data are presented in Table II.

## *Experimental*

**Solvents.** Commercially available reagent grade solvents were used without purification except as noted. 1,2,4-Trichlorobenzene was fractionally distilled and the center cut with bp 92°–93°C/16 mm was used.

**Reagents.** Substituted phenols, obtained commercially, were used unless otherwise noted; phenols were analyzed by GLC to ensure 99+% purity. 2,4,5-Trichlorophenol and pentachlorophenol were purified by sublimation and recrystallization to yield product of 99+% purity.

**Instrumentation and Standard GLC Conditions.** Preparative reactions were checked by analyzing aliquots on GLC. A Varian aerograph 1200 with flame ionization detector was used under the following conditions:

Column: 5′ × 1/8″, 5% QF-1 or OV-17
Temperatures: column—250°, injection port—300°, detector—300°
Pressures: Helium—60 psig, Hydrogen—10 psig

Quantitative determinations of purity were made at the Analytical Laboratory of Dow Chemical Co. by GLC using flame ionization and electron capture detectors. Unknown samples were compared with prepared standards containing the compounds in question.

**Preparation of the Copper (Ullmann) Catalyst.** Copper sulfate ($CuSO_4 \cdot 5H_2$, 73 grams, 0.3 mole) was weighed into a 1-liter Erlenmeyer flask and dissolved with 280 ml of deionized water. Powdered zinc was stirred into the blue solution in 0.5-gram portions until a total of 23 grams (0.35 mole) had been added. A nitrogen sparger was used to agitate the solution. Copper precipitated immediately upon addition of

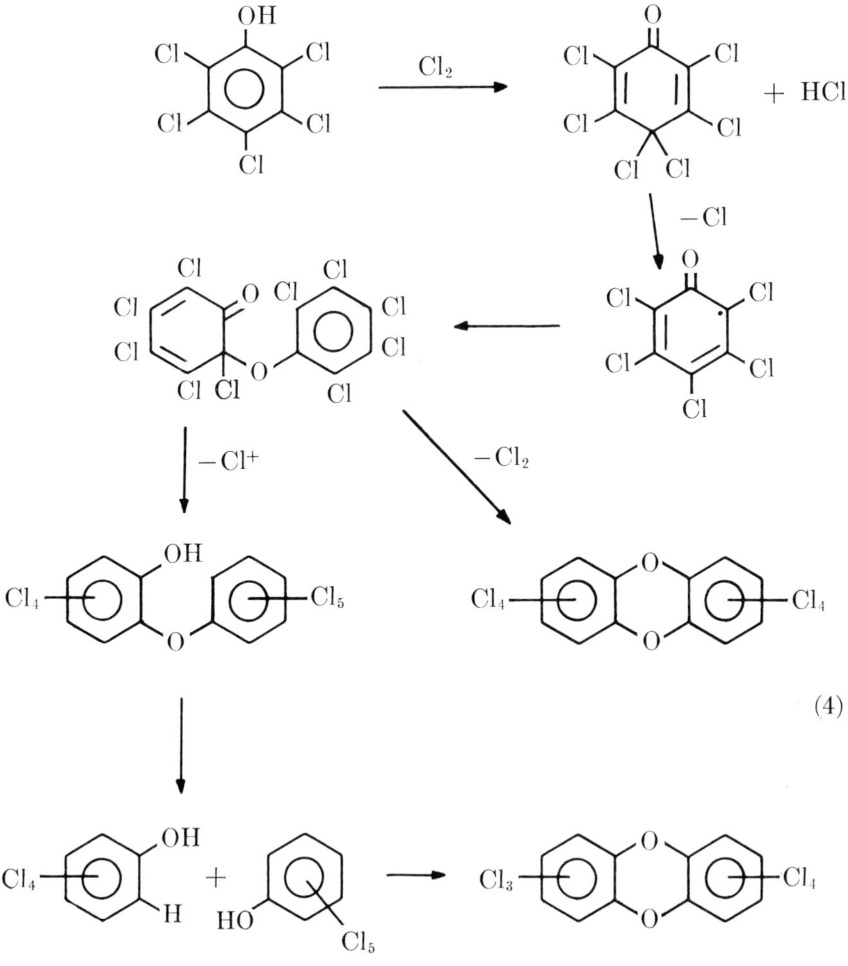

(4)

zinc powder. The flask was heated to 70°C. The mixture was decanted, and the copper residue was washed with 5% HCl solution. The precipitate was washed with water until the effluent was neutral to pH paper. Excess water was decanted and the catalyst was stored in a refrigerator. Approximately 0.5 ml of ethylene diacetate was added to the reactions with each 200 mg of catalyst.

**2-Bromo-3,4,6-trichlorophenol.** 2,4,5-Trichlorophenol (20 grams, 0.10 mole) was dissolved in 40 ml of glacial acetic acid (HOAc). Bromine (17 grams, 0.11 mole) dissolved in HOAc was added to the phenol, and the mixture was heated to 100°C. The color of bromine persisted until 30 ml of water were added. The reaction mixture yielded yellow-tinted crystals upon cooling. The product formed an oil when dissolved in methanol or HOAc–$H_2O$. The phenol was converted to its potassium salt by adding KOH and evaporating the solvent.

**Table II. Solubility of Octachlorodibenzo-*p*-dioxin**[a]

| *Solvent* | *Solubility, mgrams/liter* |
|---|---|
| Acetic acid | 48 |
| anisole | 1730 |
| chloroform | 562 |
| *o*-Dichlorobenzene | 1832 |
| dioxane | 384 |
| diphenyl oxide | 841 |
| pyridine | 400 |
| xylene | 3575 |

[a] Data from R. Papenfuss, Analytical Laboratory, The Dow Chemical Company.

**2-Bromo-4,5,6-trichlorophenol.** 2,3,4-Trichlorophenol (20 grams, 0.10 mole) was dissolved in 30 ml of HOAc and titrated with bromine (18 grams, 0.11 mole) dissolved in HOAc–$H_2O$ (30 ml each). The last few drops of bromine caused the characteristic color of bromine to persist. The product was precipitated by adding water to the reaction mixture. The filtered product was recrystallized from methanol–water solvent yielding 27 grams (98% yield) of white crystals.

**2,7-Dichlorodibenzo-*p*-dioxin.** 2-Bromo-4-chlorophenol (31 grams, 0.15 mole) and solid potassium hydroxide (8.4 grams, 0.13 mole) were dissolved in methanol and evaporated to dryness under reduced pressure. The residue was mixed with 50 ml of bEEE, 0.5 ml of ethylene diacetate, and 200 mg of copper catalyst. The turbid mixture was stirred and heated at 200°C for 15 hours. Cooling produced a thick slurry which was transferred into the 500-ml reservoir of a liquid chromatographic column (1.5 × 25 cm) packed with acetate ion exchange resin (Bio-Rad, AG1-X2, 200-400 mesh). The product was eluted from the column with 3 liters of chloroform. After evaporation, the residue was heated at 170°C/2 mm for 14 hours in a 300-cc Nestor-Faust sublimer. The identity of the sublimed product (14 grams, 74% yield) was confirmed by mass spectrometry and x-ray diffraction. Product purity was estimated at 99+% by GLC (electron capture detector).

**2,8-Dichlorodibenzo-*p*-dioxin.** 2,4,4′-Trichloro-2′-hydroxydiphenyl ether (25 grams, 86 mmole) was dissolved in methanol and mixed with an equimolar amount of potassium hydroxide. The solvent was evaporated under vacuum until the product formed a glass. The salt was dissolved in 40 ml of bEEE and the solution was heated at 200°C for 15 hours. A sample was taken from the reaction mixture, acidified with HOAc, and analyzed by GLC. Three components were observed. Under standard conditions the component eluting after 3.6 minutes represented the major product; the one eluting after 5.6 minutes represented less than 5% of the mixture, and the slowest component (6.1 minutes) had the same retention time as the starting material. Heating for 1 hour at 250° and 275°C failed to change the product distribution. The solvent was removed by distillation under reduced pressure and the residue was heated to 200°C/2 mm in a sublimer. A white powder was obtained which, when analyzed by GLC and mass spectrometry indicated the composition shown in Table III.

The dichlorodibenzo-*p*-dioxin component was isolated by passing a dioxane solution of the mixture through acetate ion exchange resin to remove phenolics. The eluted product was recrystallized from benzene. The x-ray powder diffraction pattern of the precipitate was identical with that of 2,7-dichlorodibenzo-*p*-dioxin. Analysis of the mother liquor by GLC showed a singular peak consistent with 2,7-dichlorodibenzo-*p*-dioxin. The mother liquor was cooled to 5°C and yielded transparent crystals. This material had an x-ray diffraction pattern congruent to a sample of 2,8-dichlorodibenzo-*p*-dioxin obtained from A. E. Pohland (FDA). The two patterns were quite distinct from each other (*14, 15*).

**2,3,7,8-Tetrachlorodibenzo-*p*-dioxin.** Purified 2,4,5-trichlorophenol (50 grams, 0.26 mole) was converted to its potassium salt and dissolved in 100 ml of bEEE. After addition of the copper catalyst and ethylene diacetate, the mixture was transferred to the bottom of a 300-ml sublimer with chloroform. Sublimation (200°C/2 mm) yielded 14 grams (39% yield) of 2,3,7,8-tetrachlorodibenzo-*p*-dioxin. Mass spectral analysis revealed trace quantities of pentachlorodibenzo-*p*-dioxin, tetrachlorodibenzofuran, and several unidentified substances of similar molecular weight. The combined impurity peaks were estimated to be $<1\%$ of the total integrated GLC area. The product was further purified by recrystallizations from *o*-dichlorobenzene and anisole. The final product had an estimated 260 ppm of trichlorodibenzo-*p*-dioxin as the only detected impurity.

**Hexachlorodibenzo-*p*-dioxin.** 2,3,4,6-Tetrachlorophenol was purified by distillation and recrystallization to yield a product containing $<0.1\%$ trichlorophenol impurity. The phenol was dissolved in toluene and mixed with an equimolar amount of aqueous caustic. Water was azeotropically

**Table III. GLC and Mass Spectrometry Analysis**

| *GLC Peak Area, %* | *Observed Parent Ion* m/e | *Structure* | *Molecular Weight* |
|---|---|---|---|
| 80 | 252 | 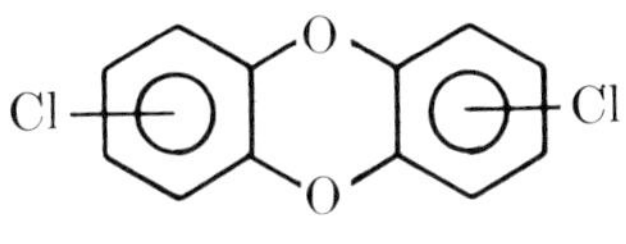 | 253 |
| 15 | 288 | 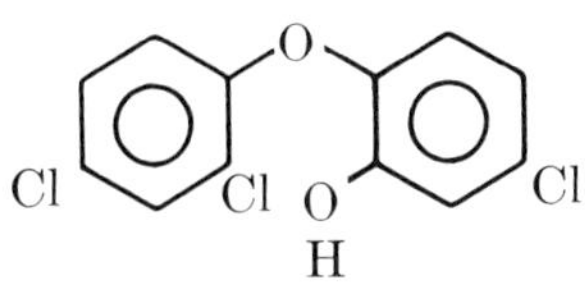 | 289.5 |
| 5 | 162 | 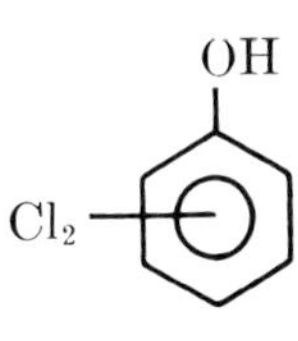 | 163 |

distilled. After trituration of the sodium 2,3,4,6-tetrachlorophenate with toluene, the salt was ground to a powder. As in a typical pyrolytic preparation, powdered sodium tetrachlorophenate (30 grams) was placed evenly on the bottom of a 300-cc Nester-Faust sublimer. The salt was covered with 20 grams of CaO and a glass wool pad. The vessel was immersed in a sand bath and heated at 350°C under reduced pressure. Temperatures over 380°C induced localized hot spots which caused smoking or bumping. Product crystals began to form on the sublimer walls after 2 hours of heating. Several product samples were isolated from their reaction mixtures by extraction with *o*-dichlorobenzene.

The *o*-dichlorobenzene extracts were combined and analyzed by GLC. Four peaks were observed under standard GLC conditions in the 10 to 15 min retention time range which is characteristic of hexachlorodibenzo-*p*-dioxins (sample 1 in Table IV). The mixture was fractionally sublimed (120° to 175°C/1 mm). The major crop was harvested at 175°C and recrystallized from anisole. Analysis of this material by GLC indicated that two isomeric hexachlorodibenzo-*p*-dioxins were present (sample 2). Overall yield (1.5 grams) of the product was 1–3% at 99+% purity, as determined by GLC and mass spectrometry.

**Table IV. GLC Analysis of *o*-Dichlorobenzene Extracts**

| *Sample* | *Description* | *Peak* | *GLC Retention Time, min* | *% of Total* |
|---|---|---|---|---|
| 1 | *o*-dichlorobenzene extract | A | 10.8 | 16 |
| | | B | 12.4 | 65 |
| | | C | 13.4 | 10 |
| | | D | 14.0 | 9 |
| 2 | sublimed and recrystallized | B | 12.0 | 55 |
| | | C | 13.5 | 45 |

**Octachlorodibenzo-*p*-dioxin.** Pentachlorophenol was purified by sublimation and recrystallization to yield a product with the following composition: trichlorophenol, 0.04%; tetrachlorophenol, 0.07%; and pentachlorophenol, 100.4 ± 1%. Pentachlorophenol (300 grams, 1.13 mole) was dissolved in 900 ml of trichlorobenzene and chlorinated anhydrously for 18 hours at reflux. Chlorine addition was stopped and the mixture was heated for 28 more hours at reflux. The crystalline product was washed with 2-liter portions of chloroform, 1*N* NaOH, methanol, and water. Analysis by GLC suggested the presence of 5–15% heptachlorodibenzo-*p*-dioxin. The mixture was carefully added to a cleaning solution of 200 ml water, 3.5 liters sulfuric acid, and 125 grams sodium dichromate. The mixture was heated at 150°C for six hours. The product was recrystallized from hot *o*-dichlorobenzene and then from anisole. The purified product (160 grams, mp 329.8° ± 0.5°C) contained <0.1% heptachlorodibenzo-*p*-dioxin, determined by GLC.

**Precautions.** Prepared materials were treated with techniques used for radioactive or infectious material. The work was performed in an isolated laboratory. Glassware and tools were segregated. The hood was lined with Saran Wrap and work areas were covered with plastic-backed paper. Waste solvents were confined and their volume was minimized, often by allowing evaporation to occur. Solid waste (towels, glassware, etc.) was placed in plastic lined containers. All wastes were burned at high temperature. Special clothing was worn when working with these chemicals and isolated when not worn. The laboratory coat was taped at the wrists and rubber gloves were worn, generally of the surgical type.

## *Literature Cited*


1. Pohland, A. E., Yang, G. C., Hansen, E. A., ADVAN. CHEM. SER. (1973) **120,** 33.
2. Tomita, M., Ueda, S., Narisada, M., *Yak. Z.* (1959) **279,** 186.
3. Hilman, H., Dietrich, J. J., *J. Org. Chem.* (1958) **22,** 1403.
4. Julia, M., Baillarge, M., *Bull. Soc. Chim. Biol.* **1953,** 640.
5. Tomita, M. *et al., Yak. Z.* (1936) **56,** 814.
6. Higginbotham, G. R., *Nature* (1968) **220,** 702.
7. Buu Hoi, N. P. *et al., C. R. Acad. Sci.* (1971) **273 (D),** 708.
8. Surrey, A. R., "Name Reactions in Organic Chemistry," Academic, New York, 1954.
9. Weingarten, H., *J. Org. Chem.* (1964) **29,** 977.
10. *Ibid.,* p. 3624.
11. Sandermann, W., Stockmann, H., Casten, R., *Chem. Ber.* (1957) **90,** 690.
12. Jones, E. L., Krizek, H., *J. Invest. Dermatol.* (1962) **39,** 511.
13. Kulka, M., *Can. J. Chem.* (1961) **31,** 1973.
14. Boer, F. P., North, P. P., *Acta Crystallogr.* (1972) **B28,** 1613.
15. Boer, F. P., Neuman, M. A., Aniline, O., *Acta Crystallogr.* (1972) **B28,** 2878.


RECEIVED August 10, 1972.

# INDEX

## E

## U



## X